T0183139

Lecture Notes in Computer Science 14540

Founding Editors

Gerhard Goos
Juris Hartmanis

Editorial Board Members

Elisa Bertino, *Purdue University, West Lafayette, IN, USA*
Wen Gao, *Peking University, Beijing, China*
Bernhard Steffen ⓘ, *TU Dortmund University, Dortmund, Germany*
Moti Yung ⓘ, *Columbia University, New York, NY, USA*

The series Lecture Notes in Computer Science (LNCS), including its subseries Lecture Notes in Artificial Intelligence (LNAI) and Lecture Notes in Bioinformatics (LNBI), has established itself as a medium for the publication of new developments in computer science and information technology research, teaching, and education.

LNCS enjoys close cooperation with the computer science R & D community, the series counts many renowned academics among its volume editors and paper authors, and collaborates with prestigious societies. Its mission is to serve this international community by providing an invaluable service, mainly focused on the publication of conference and workshop proceedings and postproceedings. LNCS commenced publication in 1973.

Nicholas Heller · Andrew Wood ·
Fabian Isensee · Tim Rädsch · Resha Teipaul ·
Nikolaos Papanikolopoulos · Christopher Weight
Editors

Kidney and Kidney Tumor Segmentation

MICCAI 2023 Challenge, KiTS 2023
Held in Conjunction with MICCAI 2023
Vancouver, BC, Canada, October 8, 2023
Proceedings

Editors
Nicholas Heller (iD)
Cleveland Clinic
Cleveland, OH, USA

Andrew Wood
Cleveland Clinic
Cleveland, OH, USA

Fabian Isensee (iD)
Helmholtz Imaging
Heidelberg, Germany

Tim Rädsch (iD)
Helmholtz Imaging
Heidelberg, Germany

Resha Teipaul (iD)
University of Minnesota
Minneapolis, MN, USA

Nikolaos Papanikolopoulos (iD)
University of Minnesota
Minneapolis, MN, USA

Christopher Weight (iD)
Cleveland Clinic
Cleveland, OH, USA

ISSN 0302-9743   ISSN 1611-3349 (electronic)
Lecture Notes in Computer Science
ISBN 978-3-031-54805-5   ISBN 978-3-031-54806-2 (eBook)
https://doi.org/10.1007/978-3-031-54806-2

© The Editor(s) (if applicable) and The Author(s), under exclusive license
to Springer Nature Switzerland AG 2024

This work is subject to copyright. All rights are reserved by the Publisher, whether the whole or part of the material is concerned, specifically the rights of translation, reprinting, reuse of illustrations, recitation, broadcasting, reproduction on microfilms or in any other physical way, and transmission or information storage and retrieval, electronic adaptation, computer software, or by similar or dissimilar methodology now known or hereafter developed.
The use of general descriptive names, registered names, trademarks, service marks, etc. in this publication does not imply, even in the absence of a specific statement, that such names are exempt from the relevant protective laws and regulations and therefore free for general use.
The publisher, the authors, and the editors are safe to assume that the advice and information in this book are believed to be true and accurate at the date of publication. Neither the publisher nor the authors or the editors give a warranty, expressed or implied, with respect to the material contained herein or for any errors or omissions that may have been made. The publisher remains neutral with regard to jurisdictional claims in published maps and institutional affiliations.

This Springer imprint is published by the registered company Springer Nature Switzerland AG
The registered company address is: Gewerbestrasse 11, 6330 Cham, Switzerland

Paper in this product is recyclable.

Preface

This volume contains the proceedings of the third international challenge on Kidney and Kidney Tumor Segmentation (KiTS 2023), held in conjunction with the 2023 International Conference on Medical Image Computing and Computer Assisted Interventions (MICCAI) in Vancouver, Canada. By "proceedings", we mean to say that this volume contains the papers written by participants in the challenge to describe their approach to developing a semantic segmentation approach for kidneys, kidney tumors, and kidney cysts, using the official training dataset released by the organizing team for this purpose.

Machine learning competitions like KiTS have become a mainstay at major machine learning and computer vision conferences, and it's hard to imagine where the field would be without them. With the unending appetite that state-of-the-art machine learning methods have for computing resources, the need for high-quality benchmarks to efficiently compare one approach to the next has never been greater. In many ways, these competitions have taken a role analogous to the cell lines and model organisms found in basic biological science research – with experiments being expensive and difficult to repeat, the need arises for a set of standardized subjects to study. For biologists, those subjects are cells and organisms. For us, those subjects are curated, high-quality test sets.

In comparison with the two KiTS challenges before it, KiTS23 featured a larger dataset and a more diverse distribution of images, and therefore a greater generalization challenge. Where KiTS19 and KiTS21 included scans only in the corticomedullary contrast phase, KiTS23 included the nephrogenic phase as well. The winning methods, as you will see, are eerily similar to those used by the winning team two and four years ago. While visual transformers are seen by many as the future of this field, they have still failed to surpass convolutional nets for this particular problem. Similarly, while large-scale pretraining has captured our imagination with generative models, the vast majority of teams still trained their models with a random initialization on the KiTS23 training set alone. We leave it for the reader to interpret whether this reflects a fundamental truth about the problem at hand or a simple practical limitation in the lack of large-scale datasets with segmented cross-sectional imaging. What is certain, however, is that large, high-quality medical datasets remain an unmet need that these competitions continue to steadily fill.

This volume includes 22 accepted papers out of 29 teams who attempted a submission to the challenge. Teams were required to submit not only a complete paper which was reviewed by at least two reviewers in a single blind manner, but also a complete set of predictions to be scored against the private test set. Our sincere gratitude is offered to those who participated in this competition. Kidney cancer does not represent the largest share of either new cancer diagnoses or cancer deaths, but it is a substantial problem nonetheless, affecting hundreds of thousands of individuals around the world. Through collaborations like KiTS, we have made major strides in automatically segmenting these tumors, and subsequently in understanding how tumor morphology relates to natural history and treatment outcomes.

December 2023

Nicholas Heller
Andrew Wood
Fabian Isensee
Tim Rädsch
Resha Tejpaul
Nikolaos Papanikolopoulos
Christopher Weight

Organization

Organizing Committee

Nicholas Heller	Cleveland Clinic, USA
Andrew Wood	Cleveland Clinic, USA
Fabian Isensee	Helmholtz Imaging, Germany
Tim Rädsch	Helmholtz Imaging, Germany
Resha Tejpaul	University of Minnesota, USA
Nikolaos Papanikolopoulos	University of Minnesota, USA
Christopher Weight	Cleveland Clinic, USA

Program Committee

Rebecca Campbell	Cleveland Clinic, USA
Onurlap Ergun	University of Minnesota, USA
Nour Abdallah	Cleveland Clinic, USA
Ghady Zgheib	Saint Joseph University of Beirut, Lebanon
Caleb Curry	Case Western Reserve University, USA
Kyle Harris	Cleveland Clinic, USA
Alex You	Cleveland Clinic, USA

Contents

Automated 3D Segmentation of Kidneys and Tumors in MICCAI KiTS 2023 Challenge

Andriy Myronenko$^{(\boxtimes)}$, Dong Yang, Yufan He, and Daguang Xu

NVIDIA, Santa Clara, USA
amyronenko@nvidia.com

Abstract. Kidney and Kidney Tumor Segmentation Challenge (KiTS) 2023 [6] offers a platform for researchers to compare their solutions to segmentation from 3D CT. In this work, we describe our submission to the challenge using automated segmentation of Auto3DSeg (https://monai.io/apps/auto3dseg) available in MONAI (https://github.com/Project-MONAI/MONAI). Our solution achieves the average dice of 0.835 and surface dice of 0.723, which ranks first and wins the KiTS 2023 challenge (https://kits-challenge.org/kits23/#kits23-official-results).

Keywords: Auto3DSeg · MONAI · Segmentation

1 Introduction

Almost half a million people are diagnosed with kidney cancer annually. Each year, a larger number of kidney tumors are detected, and currently, it is difficult to determine whether a tumor is malignant or benign using radiographic methods. The risk of metastatic progression remains a serious concern, highlighting the need for reliable systems to objectively characterize kidney tumor images and predict treatment outcomes.

For almost five years, the KiTS [5] initiative has maintained and expanded a publicly available collection of hundreds of segmented CT scans featuring kidney tumors. This year's KiTS'23 [6] competition includes an expanded training set consisting of 489 cases. The goal of the challenge is to develop an automated method to segment kidneys, tumors and cysts.

2 Methods

We implemented our approach with MONAI [1] using Auto3DSeg open source project. Auto3DSeg is an automated solution for 3D medical image segmentation, utilizing open source components in MONAI, offering both beginner and advanced researchers the means to effectively develop and deploy high-performing segmentation algorithms.

The minimal user input to run Auto3DSeg for KiTS'23 datasets, is

ⓒ The Author(s), under exclusive license to Springer Nature Switzerland AG 2024
N. Heller et al. (Eds.): KiTS 2023, LNCS 14540, pp. 1–7, 2024.
https://doi.org/10.1007/978-3-031-54806-2_1

```bash
#!/bin/bash
python -m monai.apps.auto3dseg AutoRunner run \
    --input="./input.yaml"
```

where a user provided input config (input.yaml) includes only a few lines:

```yaml
# This is the YAML file "input.yaml"
modality: CT
datalist: "./dataset.json"
dataroot: "/data/kits23"

class_names:
- { "name": "kidney_and_mass", "index": [1,2,3] }
- { "name": "mass", "index": [2,3] }
- { "name": "tumor", "index": [2] }
sigmoid : true
```

When running this command, Auto3DSeg will analyze the dataset, generate hyperparameter configurations for several supported algorithms, train them, and produce inference and ensembling. The system will automatically scale to all available GPUs and also supports multi-node training.

The 3 minimum user options (in input.yaml) are data modality (CT in this case), location of the downloaded KiTS'23 dataset (dataroot), and the list of input filenames with an associated fold number (dataset.json). We generate the 5-fold split assignments randomly. Since KiTS defines its specific label mapping (from integer class labels to 3 subregions, see Fig. 1), we have to define it in the config, and since these subregions are overlapping, we use "sigmoid: true" to indicate multi-label segmentation, where the final activation is sigmoid (instead of the default softmax).

Currently, the default Auto3DSeg setting trains three 3D segmentation algorithms: SegResNet [8], DiNTS [4] and SwinUNETR [3,9] with their unique training recipes. SegResNet and DiNTS are convolutional neural network (CNN) based architectures, whereas SwinUNETR is based on transformers. Each is trained using 5-fold cross validation.

For model inference, a sliding-window scheme is used to create probability maps, which are re-sampled back to its original spacing. This allows ensembling prediction from different algorithms even if there were trained at different resolutions.

The simplicity of Auto3DSeg is a very minimal user input, which allows even non-expert users to achieve a great baseline performance. The system will take care of most of the heavy lifting to analyze, configure and optimally utilize the available GPU resources. And for expert users, there are many configuration options that can be manually provided to override the automatic values, for better performance tuning.

In the final prediction, we ensemble the best model checkpoints only from SegResNet and DiNTS algorithms, since they performed better during

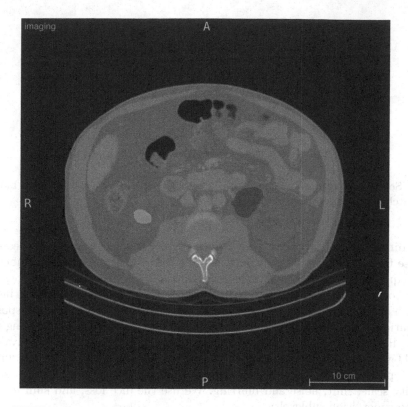

Fig. 1. KiTS'23 data example of an axial slice with the provided annotations of kidneys (red), tumors (green) and cysts (blue). The classes of interest that KiTS tasks to segment are: a) all foreground combined b) tumors + cysts (green and blue) c) tumors only (green). (Color figure online)

cross-validation. We also applied a few small customizations to the baseline Auto3DSeg workflow. We describe the baseline Auto3DSeg method and the customization below.

2.1 Training and Validation Data

Our submission made use of the official KiTS'23 training set alone.

2.2 SegResNet

SegResNet[1] is an encode-decoder based semantic segmentation network based on [8]. It is a U-net based convolutional neural network with deep supervision (see Fig. 2).

The default Auto3DSeg SegResNet configuration was used, which includes 5 levels of 1, 2, 2, 4, 4 blocks. It follows a common CNN approach to downsize

---

[1] https://docs.monai.io/en/stable/networks.html.

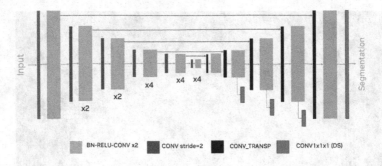

Fig. 2. SegResNet network configuration. The network uses repeated ResNet blocks with batch normalization and deep supervision

image dimensions by 2 progressively (down to 16× smaller) and simultaneously increase feature size by. All convolutions are $3 \times 3 \times 3$ with an initial number of filters equal to 32. The encoder is trained with a $256 \times 256 \times 256$ input region. The decoder structure is similar to the encoder one, but with a single block per each spatial level. Each decoder level begins with upsizing with transposed convolution: reducing the number of features by a factor of 2 and doubling the spatial dimension, followed by the addition of encoder output of the equivalent spatial level. The number of levels and the region size is automatically configured. We use spatial augmentation including random affine and flip in all axes, random intensity scale, shift, noise and blurring. We use the dice loss, and sum it over all deep-supervision sublevels:

$$Loss = \sum_{i=0}^{4} \frac{1}{2^i} Loss(pred, target^{\downarrow}) \tag{1}$$

where the weight $\frac{1}{2^i}$ is smaller for each sublevel (smaller image size) i. The target labels are downsized (if necessary) to match the corresponding output size using nearest neighbor interpolation.

We use the AdamW optimizer with an initial learning rate of $2e^{-4}$ and decrease it to zero at the end of the final epoch using the Cosine annealing scheduler. We use batch size of 1 per GPU, We use weight decay regularization of $1e^{-5}$. Input images were re-scaled from $[-54, 242]$ to the $[-1, 1]$ CT interval, followed by a sigmoid function. The range was determined automatically by the data analysis step to include the intensity pattern variations within the foreground regions.

2.3 DiNTS

DiNTS stands for Differentiable Network Topology Search (DiNTS) scheme, an advanced methodology that fosters more dynamic topologies and an integrated two-level search. DiNTS has demonstrated superior performance, achieving top-tier results in the Medical Segmentation Decathlon (MSD) challenge [2].

The DiNTS algorithm utilizes a densely-connected lattice-based network, training with a $96 \times 96 \times 96$ model input for both training and inference. It leverages automatic mixed precision (AMP) and the SGD optimizer, with an initial learning rate of 0.2 and a loss defined by the Dice plus focal Loss. For the data processing, we utilize intensity normalization and random cropping, as well as random rotation, zoom, Gaussian smoothing, intensity scaling and shifting, Gaussian noising, and flipping.

In the quest for enhanced model performance, we fine-tuned the checkpoints that were initially trained using the default training recipe, adjusting them with various patch sizes for a span of 25 epochs. Our observations suggest that a larger patch size often leads to improved model performance. Taking into account our computational budget, we have selected a patch size from a range between 192^3 and $192 \times 192 \times 288$ for each fold of the model (based on validation Dice scores) as the configuration in our final model inference.

2.4 Metrics

The output of the network has 3 channels followed by a sigmoid, to segment each of the 3 KiTS'23 expected classes: a) Kidney + Tumor + Cyst; b) Tumor + Cyst; c) Tumor only. This creates a multi-label segmentation, where each voxel can belong to more than one label. We use an average dice metric of these 3 classes to select the best validation checkpoints (without considering surface distances).

2.5 Auto3DSeg Customizations for KiTS'23

Even though the default Auto3DSeg configuration achieved a good baseline cross-validation performance automatically, we did a few customization including cropping to kidneys region and post-processing.

Training on the full size 3D CT images can be time consuming, so we pre-cropped the images around the kidneys region. A simple rectangular box was used, based on the ground truth labels. Since we used only 1 cropping per image, the cropped region included not only kidneys but everything in between including the spine. Training on such cropped images has 2 advantages: firstly, it allowed for faster training, since smaller images can be cached in RAM, and secondly, it simplified the task for the network. The disadvantage of such an approach is that it requires finding the bounding box of the kidneys region first.

We trained a separate segmentation network to find the foreground and calculate the kidneys bounding box coordinates. For all the tasks we used the same exact network architecture, trained all at $0.78 \times 0.78 \times 0.78 \, \text{mm}^3$ CT resolution. Arguably, bounding box detection could have been done faster, using a simpler detection network and at a lower CT resolution, but here we saved on coding time, by reusing the framework. We trained the first round of models fast (using a smaller number of epochs), to be used as a bounding box detector. And after that trained longer, using only the cropped (around kidneys) regions. This approach is somewhat similar in spirit to the KiTS 2021 champion solution of coarse-to-fine training [10] (based on nnU-net [7]), but here we do not re-use or concatenate masks detected at a coarse level, we simply use it to detect the bounding box for faster training.

We also added binary post-processing on the final segmentation masks. Firstly, we remove small connected components (smaller then 100 voxels total) based on the foreground (merged labels). Secondly, we correct for "outline" of some tumor region. During a training stage, we noticed that on a small set of images, network predictions of the tumor label have a small rim (1–2 voxels) of cyst label. This happened mostly because the network was trained as a multi-label task, where each voxel can be assigned to several classes. Since, it's not possible for a tumor to have a "cystic" outline (even if it looks like a cyst image pattern), by definition, we decided to correct such cases with a simple binary post-processing. In our cross-validation tests, this final post-processing did not actually affect the accuracy metrics, but we still decided to include it.

Finally, we increased the size of the network input patch during training to $256 \times 256 \times 256$ for SegResNet and to $192 \times 192 \times 288$ for DiNTS, which allowed for faster training and also slightly increased the cross-validation performance.

2.6   Optimization

We train the method on an 8-GPU 48 GB NVIDIA A40 machine, with a batch size of 1 per GPU, which is equivalent to batch size of 8 single GPU training. Auto3DSeg caches on-the-fly all the resampled data in RAM during the first training epoch, when sufficient amount of RAM is available (otherwise a fraction of the data is cached). This way only the first epoch suffers a slow-down due to disk i/o and resampling, and the rest of the training process is fast.

3   Results

Based on our random 5-fold split, the average dice scores per fold are shown in Table 1. For the final submission we used an ensemble of 15 models: 10 models of SegResNet (5 folds trained twice), and 5 models of DiNTS. SegResNet A and B training runs in Table 1 had the same configurations.

Table 1. Average Dice results of the 15 trained models based on our 5-fold data split.

	Fold 1	Fold 2	Fold 3	Fold 4	Fold 5	Average
SegResNet A	0.8997	0.8739	0.8923	0.8911	0.8892	0.88924
SegResNet B	0.8995	0.8773	0.8913	0.889	0.8865	0.88872
DiNTS	0.8810	0.8647	0.8806	0.8752	0.8822	0.8767

On the final hidden challenge dataset, our submission achieved an average Dice score of 0.835, which ranked first among other submissions.

4   Conclusion

We described our winning solution to KiTS 2023 challenge using Auto3DSeg from MONAI. Our final submission is en ensemble of 15 CNN models, 10 of SegResNet and 5 of DiNTS. We hope that open source tools in MONAI will help more researchers to achieve good baseline 3D segmentation results on their particular task. Our solution achieves the average dice of 0.835 and surface dice of 0.723, which ranks first on the KiTS 2023 leaderboard[2].

References

1. Project-MONAI/MONAI. https://doi.org/10.5281/zenodo.5083813
2. Antonelli, M., et al.: The medical segmentation decathlon. Nat. Commun. **13**(1), 4128 (2022)
3. Hatamizadeh, A., Nath, V., Tang, Y., Yang, D., Roth, H.R., Xu, D.: Swin UNETR: swin transformers for semantic segmentation of brain tumors in MRI images. In: Crimi, A., Bakas, S. (eds.) BrainLes 2021. LNCS, vol. 12962, pp. 272–284. Springer, Cham (2022). https://doi.org/10.1007/978-3-031-08999-2_22
4. He, Y., Yang, D., Roth, H., Zhao, C., Xu, D.: Dints: differentiable neural network topology search for 3D medical image segmentation. In: Proceedings of the IEEE/CVF Conference on Computer Vision and Pattern Recognition, pp. 5841–5850 (2021)
5. Heller, N., et al.: The state of the art in kidney and kidney tumor segmentation in contrast-enhanced CT imaging: results of the KiTS19 challenge. Med. Image Anal. **67**, 101821 (2021)
6. Heller, N., et al.: The 2023 kidney and kidney tumor segmentation challenge. https://kits-challenge.org/kits23/
7. Isensee, F., Jaeger, P.F., Kohl, S.A.A., Petersen, J., Maier-Hein, K.H.: nnU-Net: a self-configuring method for deep learning-based biomedical image segmentation. Nat. Methods **18**, 203–211 (2021)
8. Myronenko, A.: 3D MRI brain tumor segmentation using autoencoder regularization. In: Crimi, A., Bakas, S., Kuijf, H., Keyvan, F., Reyes, M., van Walsum, T. (eds.) BrainLes 2018. LNCS, vol. 11384, pp. 311–320. Springer, Cham (2019). https://doi.org/10.1007/978-3-030-11726-9_28
9. Tang, Y., et al.: Self-supervised pre-training of swin transformers for 3D medical image analysis. In: Proceedings of the IEEE/CVF Conference on Computer Vision and Pattern Recognition, pp. 20730–20740 (2022)
10. Zhao, Z., Chen, H., Wang, L.: A coarse-to-fine framework for the 2021 kidney and kidney tumor segmentation challenge. In: Heller, N., Isensee, F., Trofimova, D., Tejpaul, R., Papanikolopoulos, N., Weight, C. (eds.) Kidney and Kidney Tumor Segmentation, pp. 53–58 (2022)

---

[2] https://kits-challenge.org/kits23/#kits23-official-results.

Exploring 3D U-Net Training Configurations and Post-processing Strategies for the MICCAI 2023 Kidney and Tumor Segmentation Challenge

Kwang-Hyun Uhm[1], Hyunjun Cho[1], Zhixin Xu[1], Seohoon Lim[1], Seung-Won Jung[1], Sung-Hoo Hong[2], and Sung-Jea Ko[1,3(✉)]

[1] Korea University, Seoul, South Korea
khuhm@dali.korea.ac.kr, sjko@korea.ac.kr
[2] The Catholic University of Korea, Seoul, South Korea
[3] MedAI, Seoul, South Korea

Abstract. In 2023, it is estimated that 81,800 kidney cancer cases will be newly diagnosed, and 14,890 people will die from this cancer in the United States. Preoperative dynamic contrast-enhanced abdominal computed tomography (CT) is often used for detecting lesions. However, there exists inter-observer variability due to subtle differences in the imaging features of kidney and kidney tumors. In this paper, we explore various 3D U-Net training configurations and effective post-processing strategies for accurate segmentation of kidneys, cysts, and kidney tumors in CT images. We validated our model on the dataset of the 2023 Kidney and Kidney Tumor Segmentation (KiTS23) challenge. Our method took the second place in the final ranking of KiTS23 challenge on unseen test data with an average Dice score of 0.820 and an average Surface Dice of 0.712.

Keywords: Kidney cancer · Medical image segmentation · 3D U-Net

1 Introduction

In 2023, it is estimated that 81,800 kidney cancer cases will be newly diagnosed, and 14,890 people will die from this cancer in the United States [1]. Kidney cancer is one of the 10 most common cancers, and by far the most common type of kidney cancer is renal cell carcinoma (RCC), which occurs in 9 out of 10 cases of all kidney cancer [7]. Preoperative dynamic contrast-enhanced abdominal computed tomography (CT) is often used for the detection and evaluation of renal tumors [6]. However, there are some overlaps in image-level features between kidneys, cysts, and renal tumors, which make accurate segmentation difficult and cause inter-observer variation. These clinical issues point to the need to develop automatic systems that can reduce misdiagnosis and inter-observer variation.

In this paper, we explore various 3D U-Net training configurations and effective post-processing strategies for accurate segmentation of kidneys, cysts, and

ⓒ The Author(s), under exclusive license to Springer Nature Switzerland AG 2024
N. Heller et al. (Eds.): KiTS 2023, LNCS 14540, pp. 8–13, 2024.
https://doi.org/10.1007/978-3-031-54806-2_2

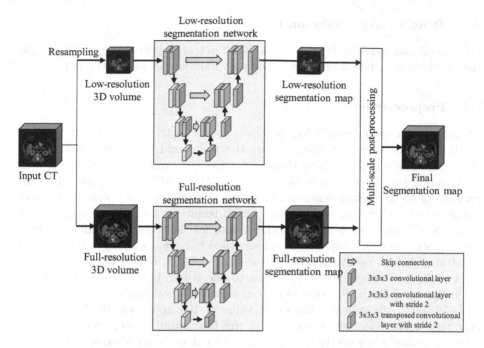

Fig. 1. Overview of our multi-scale prediction fusion framework.

kidney tumors in CT images. We investigate a wide variety of training configurations including training at different scales, cascade training approaches, and region-based training. We also introduce post-processing approaches which aim at improving the performance by effectively combining the predictions from the models trained in different training configurations. We validated our model on the dataset of 2023 Kidney and Kidney Tumor Segmentation (KiTS23) challenge, which includes data from previous challenges, *e.g.*, KiTS19 [3].

2   Methods

In this paper, we propose to perform segmentation at both low-resolution and full-resolution and then combine those two predictions based on a task-specific post-processing scheme, as shown in Fig. 1. Low-resolution 3D CT volumes are generated by resampling original input CT images. Two independent 3D U-Nets are utilized to produce low-resolution segmentation maps and full-resolution segmentation maps from low-resolution 3D CT volume and full-resolution CT volume, respectively. The final segmentation map is produced by multi-scale post-processing which takes the predictions of low- and full-resolution segmentation maps and combines them with domain-specific procedures.

2.1   Training and Validation Data

Our submission made use of the official KiTS23 training set alone [4]. We divide the provided data into a training set and a validation set at a ratio of 4:1.

2.2   Preprocessing

We follow the way in the nnU-Net [5] to preprocess the training data. The spacing of all official CT images is the same on the x-axis and y-axis, but different on the z-axis. The original training data have a median voxel spacing of $0.78 \times 0.78 \times 3.0\,mm^3$, and the median volume shape of $512 \times 512 \times 104$ voxels. For the training of low-resolution segmentation network, input data is resampled to have a spacing of $1.84 \times 1.84 \times 2.36\,mm^3$, which results in a median volume of $217 \times 217 \times 177\,mm^3$ voxels. On the other hand, for the training of full-resolution segmentation network, input data is resampled to have a spacing of $0.78 \times 0.78 \times 1.0\,mm^3$, which results in a median volume of $512 \times 512 \times 417\,mm^3$ voxels.

We clip each case's intensity values to the 0.5 and 99.5 percentiles of the intensity values in the foreground regions across the training set, *i.e.*, the range of $[-58, 302]$. We subtract the mean value of 103 and then divide it by the standard deviation of the intensities in the foreground regions, which is 73.3. The foreground class oversampling is used to enforce more than a third of the samples in a batch contain at least one randomly chosen foreground class. During training, patches with shape $128 \times 128 \times 128$ are sampled and input to the network. A variety of data augmentation techniques are applied on the fly during training: rotations, scaling, mirroring, etc.

2.3   Proposed Method

Network Architecture. We use 3D U-Net architecture [2] for both low-resolution and full-resolution segmentation networks. The U-Net consists of an encoder and decoder, where for all convolutions in the network we use $3 \times 3 \times 3$ convolution kernels. Each block in the encoder consists of a sequence of two convolutional layers each of which followed by instance normalization and LeakyReLU activations. In the decoder, upsampling is done by $3 \times 3 \times 3$ transposed convolutions. At the last convolutional layer, it outputs the probability distributions for background, kidney, cyst, and tumor for each voxel. There are a total of 6 stages for each encoder and decoder. We use cross-entropy loss and dice loss for the training. The stochastic gradient descent strategy is used for the optimization. We investigate various network configurations. For example, we increase all the channel numbers of convolutions in the network by two times. This increases the model parameter by two times, and also the training time. Another variant is a residual 3D U-Net which replaces plain convolution blocks in the encoder with residual blocks. Also, we tested the region-based training, which uses three sigmoid activations after the final convolutional layer to produce probabilities for each region, where regions are defined by "kidney and masses",

Table 1. Results of experiments on the validation set.

Method	Kidney	Masses	Tumor	Average
Low-resolution	0.973	**0.858**	0.794	0.875
Low-resolution - channel × 2	0.973	0.851	0.771	0.865
Low-resolution - residual	0.973	0.856	0.794	0.874
Full-resolution	0.977	0.840	0.790	0.869
Full-resolution - channel × 2	0.975	0.840	0.790	0.868
Full-resolution - residual	**0.979**	**0.858**	0.803	0.880
Full-resolution - batch 4	0.978	0.857	0.801	0.879
Cascade	**0.979**	**0.858**	0.804	0.880
Region-based training	0.975	0.851	0.790	0.872
Ours (w/ post-processing)	**0.979**	0.857	**0.826**	**0.887**

"masses", and "tumor". For the cascade model, the output of the low-resolution segmentation network is concatenated to the input image and then fed to the high-resolution segmentation network.

Multi-scale Post-processing. Based on our analysis of the results of the validation set, the low-resolution segmentation network produces well-localized segmentation parts but lacks sufficient details, while the full-resolution segmentation model provides finely detailed boundaries of kidneys and masses but generates some false positives around backgrounds. Therefore, we remove segmented foreground blobs in full-resolution segmentation maps which do not belong to the foreground blobs in low-resolution segmentation maps. Moreover, to reduce tumor false positives (FPs), we first perform the connected component analysis for the tumor class, and then we treat the tumor regions in each segmentation map as true positives if they have sufficient overlap with the tumor regions of the segmentation map from another scale. Specifically, the two tumor regions from low- and full-resolution segmentation maps should have a Dice coefficient greater than 0.3 to be determined as true tumor regions. Finally, we join the regions of predicted segmentation parts from both the full-resolution and low-resolution segmentation maps to complement each other. This enables us to take advantage of both the low-resolution and high-resolution segmentation predictions, boosting the final segmentation performance. In addition, we find the convex hulls of tumor blobs in the predicted segmentation maps and then merge the labels inside the detected hulls to reduce noisy prediction results. We remove foreground blobs that have an area smaller than $10{,}000\,\mathrm{mm}^3$.

3   Results

We validated our model on the dataset of the 2023 Kidney and Kidney Tumor Segmentation (KiTS23) challenge. We report performances of baselines and our

Table 2. KiTS23 leaderboard for final results on test data (top-5).

Rank	Team	Affiliation	Dice	Surface Dice	Tumor Dice
1	Andriy Myronenko et al.	NVIDIA	0.835	0.723	0.756
2	Kwang-Hyun Uhm	Korea University	0.820	0.712	0.738
3	Yasmeen George	Monash University	0.819	0.707	0.713
4	Shuolin Liu	Independent Researcher	0.805	0.706	0.697
5	George Stoica et al.	University of Lasi, SenticLab	0.807	0.691	0.713

models evaluated on the validation set. We report Dice scores for regions of "Kidney and Masses", "Kidney Masses", and "Kidney Tumor". For brevity, we denote in the tables in this section the dice scores as "Kidney", "Masses", and "Tumor".

We summarize the quantitative results in Table 1. All the results are based on the validation set, which contains 98 cases. We can see that our method outperforms other baselines by a large margin in terms of average Dice score. The average Dice is 0.887, and Dice for kidney, kidney masses, and kidney tumors are respectively 0.979, 0.857, and 0.826. For the tumor segmentation, our algorithm performs significantly better than the baseline. These results demonstrate the effectiveness of our multi-scale post-processing strategy.

Our final submission involves three base networks including "Low-resolution", "Low-resolution - residual", and "Full-resolution - batch 4" which show better performance than others in several folds. We apply our multi-scale post-processing strategy to two possible pairs of low- and full-resolution segmentation results and join the two post-processed results to make the final segmentation map (Table 2).

4   Discussion and Conclusion

In this paper, we explore various 3D U-Net training configurations and effective post-processing strategies for accurate segmentation of kidneys, cysts, and kidney tumors in CT images. We investigate a wide variety of training configurations including training at different scales, cascade training approaches, and region-based training. We also introduce post-processing approaches that aim at improving performance by effectively combining the predictions from the models trained in different training configurations. We validated our model on the dataset of 2023 Kidney and Kidney Tumor Segmentation (KiTS23) challenge.

Our approach won second place in KiTS23 Challenge. On the test set, our final model obtained Dice scores of 0.948, 0.776, and 0.738 and Surface Dice (SD) scores of 0.899, 0.635 and 0.602 for kidney, masses, and tumors, respectively. It is important to note that by carefully combining the results of 3D U-Nets trained from different scale settings, we can obtain much better performance than the results of individual models. Specifically, we achieved a much higher tumor dice score than lower-ranked teams, which demonstrates that our multi-scale tumor prediction aggregation strategy is effective for accurate tumor segmentation. Our method can be applied to other multi-scale approaches in current literature to improve the segmentation performance for accurate diagnosis. Our work is only validated on KiTS23 Challenge dataset and lacks sufficient experimental validation on multiple datasets. In the future, we will expand the experiment settings of our method to different datasets for comprehensive evaluation of generality.

Acknowledgment. We thank the KiTS competition organizers, data providers, and annotators for their great effort in the challenge. We further thank the creator and contributors to the nnU-Net framework.

This work was supported by the Korea Medical Device Development Fund grant funded by the Korea government (the Ministry of Science and ICT, the Ministry of Trade, Industry and Energy, the Ministry of Health & Welfare, the Ministry of Food and Drug Safety) (Project Number: 1711195432, RS-2020-KD000096).

References

1. American cancer society. About kidney cancer. https://www.cancer.org/cancer/kidney-cancer/about.html. Accessed 14 Aug 2023
2. Çiçek, Ö., Abdulkadir, A., Lienkamp, S.S., Brox, T., Ronneberger, O.: 3D U-Net: learning dense volumetric segmentation from sparse annotation. In: Ourselin, S., Joskowicz, L., Sabuncu, M.R., Unal, G., Wells, W. (eds.) MICCAI 2016. LNCS, vol. 9901, pp. 424–432. Springer, Cham (2016). https://doi.org/10.1007/978-3-319-46723-8_49
3. Heller, N., et al.: The state of the art in kidney and kidney tumor segmentation in contrast-enhanced CT imaging: results of the KiTS19 challenge. Med. Image Anal. **67**, 101821 (2021)
4. Heller, N., et al.: The KiTS19 challenge data: 300 kidney tumor cases with clinical context, CT semantic segmentations, and surgical outcomes. arXiv preprint arXiv:1904.00445 (2019)
5. Isensee, F., et al.: nnU-Net: self-adapting framework for u-net-based medical image segmentation. CoRR abs/1809.10486 (2018). http://arxiv.org/abs/1809.10486
6. Uhm, K.H., Jung, S.W., Choi, M.H., Hong, S.H., Ko, S.J.: A unified multi-phase CT synthesis and classification framework for kidney cancer diagnosis with incomplete data. IEEE J. Biomed. Health Inform. **26**(12), 6093–6104 (2022)
7. Uhm, K.H., et al.: Deep learning for end-to-end kidney cancer diagnosis on multiphase abdominal computed tomography. NPJ Precis. Oncol. **5**(54) (2021)

Dynamic Resolution Network for Kidney Tumor Segmentation

Shuolin Liu[✉] and Bing Han

Canon Medical Systems (China) Co., Ltd., Beijing, China
shuolin1.liu@cn.medical.canon

Abstract. Segmentation of kidneys, kidney tumors and kidney cysts from contrast-enhanced CT images has significant potential to facilitate large-scale imaging and radiological analysis. However, the task is challenging due to the considerable variation in tumor scales between different cases, which is not effectively addressed by conventional segmentation methods. In this paper, we propose a method called dynamic resolution that addresses this issue by dynamically adjusting the image resolution for each sample during training and testing, thus achieving a balance between targets with different scales. We also present a technique that uses publicly available unlabelled datasets to improve the robustness of the model without requiring additional manual labelling. We evaluated our method on the KiTS23 competition dataset and the results demonstrate its superiority over the existing state-of-the-art nnUnet, with improvements of 1.2%, 3.9% and 4.9% on kidney, tumor+cyst and tumor respectively.

Keywords: Kidney tumor segmentation · Dynamic resolution · Semi-supervised learning

1 Introduction

High-quality semantic segmentation of the kidneys, kidney tumors and kidney cysts from contrast-enhanced CT images has enormous potential for large-scale radiological analysis of kidney tumor imaging and its correlation with tumor molecular features and disease-specific outcomes [2]. Unfortunately, manual semantic segmentation of these structures is very time-consuming in routine clinical practice. There is still an unmet need for highly accurate and versatile automated segmentation of these structures, which can significantly reduce the repetitive manual work for human during the analysis process.

Deep learning methods, particularly fully convolutional neural networks (FCNs) [6], have emerged as powerful tools in the field of medical image analysis [3–5]. Among the various approaches, U-Net [5] and its variants have demonstrated exceptional success in segmenting anatomical structures such as organs and lesions. Notably, nnUnet [3] stands out as one of the most effective variations of U-Net, as it incorporates a self-configuring method that enhances the

© The Author(s), under exclusive license to Springer Nature Switzerland AG 2024
N. Heller et al. (Eds.): KiTS 2023, LNCS 14540, pp. 14–21, 2024.
https://doi.org/10.1007/978-3-031-54806-2_3

adaptability of the algorithm to diverse datasets and tasks, resulting in improved versatility and high accuracy. Despite these advancements, there are still persistent challenges that need to be addressed. Specifically, existing methods struggle with preserving fine structures and effectively handling significant variations in the scale of the target being segmented. We argue that a limitation of current methods lies in their handling of multi-scale targets, where the same parameters are applied indiscriminately to targets of different scales. As shown in Fig. 1, when identical patch sizes and resolutions are used to process tumor A and tumor B, the field of view (FOV) for tumor A is too small, while that for tumor B is too large. Consequently, this discrepancy affects the accuracy of the final segmentation results.

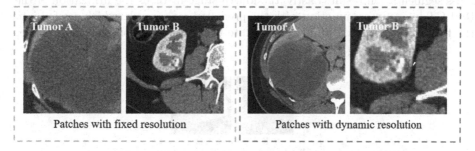

Patches with fixed resolution   Patches with dynamic resolution

Fig. 1. Examples of patches used for training and testing. From left to right are patches obtained using the fixed resolution and dynamic resolution methods.

In this study, we present a novel approach, called dynamic resolution, to address above challenges associated with medical image analysis. Inspired by the practices of medical professionals, our method (called DR-Net hereafter)involves dynamically adjusting the field of view (FOV) based on the size of the target being observed. Specifically, the FOV is increased for larger targets and decreased for smaller ones. By incorporating dynamic resolution, the DR-Net enables adaptive resolution adjustment for all targets during both training and testing phases. This adaptability ensures the selection of an appropriate FOV size, ultimately improving the accuracy of segmentation. Furthermore, we propose the integration of exclusion learning into the task of kidney tumor segmentation. This technique enhances the robustness of our model by leveraging other publicly available CT image datasets, even in the absence of kidney tumor annotations. Our approach offers a promising solution to improve the performance of kidney tumor segmentation in clinical practice.

We conducted a comprehensive evaluation of the proposed method on the KiTS2023[1] competition dataset. Our results demonstrate that our approach outperforms the current state-of-the-art method.

---

[1] https://kits-challenge.org/kits23/.

2   Method

Figure 2 illustrates the comprehensive framework of DR-Net, which is developed for the purpose of kidney tumor segmentation. In kidney tumor segmentation, the background region of the image typically occupies the majority of the image. To enhance the framework's speed and robustness, inspired by Zhao et al. [8], DR-Net is constructed in four distinct stages. The first stage involves ROI localization, where a low-resolution model is employed to detect the Region of Interest (ROI) within the CT image. The identified ROI is then forwarded to the Stage 2 model, which generates precise kidney segmentation results. Moving forward, the third stage encompasses mass segmentation, which takes the ROI CT image and the predicted kidney segmentation outcome as inputs, and produces the mass segmentation result. Lastly, the fourth stage involves the utilization of the ROI CT image in conjunction with the predicted mass mask, which are fed into the tumor segmentation model to generate the final tumor prediction outcome.

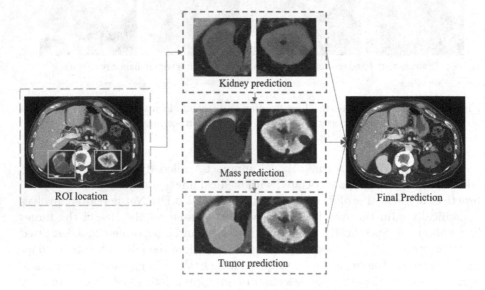

Fig. 2. Overall framework of DR-Net. "Mass" indicates tumor+cyst.

2.1   Dynamic Resolution for Medical Segmentation

To address the variations in sizes and shapes of different targets, dynamic resolution was employed for stage2 and stage3. This approach emulates the segmentation process performed by human doctors when analyzing medical images. Specifically, for large targets, they are labeled at a regular scale, whereas for

Table 1. Model parameters used for each stage.

Stage	Resolution(mm)	Patch size
ROI location	(3.0, 1.5, 1.5)	(96, 192, 192)
Kidney segmentation	Dynamic	(160, 128, 128)
Mass segmentation	Dynamic	(128, 128, 128)
Tumor segmentation	(0.78, 0.78, 0.78)	(128, 128, 128

small organs, a two-step approach is adopted. First, the small organs are localized, and then a zoom-in technique is applied to achieve more precise segmentation (Fig. 3).

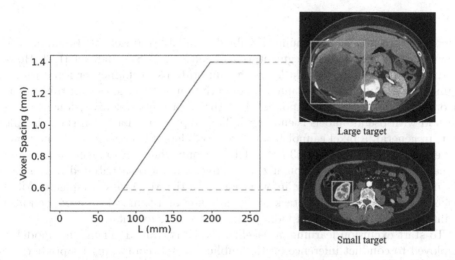

Fig. 3. The illustration of dynamic resolution.

In the task of kidney tumor segmentation, our findings indicate that when the tumor size is excessively large, it leads to an increase in the overall kidney volume. Consequently, we have chosen the length of the kidney bounding box (bbox) as the criterion for assessing the size of the target. For medical images, modifying the image resolution is equivalent to adjusting the voxel spacing. To determine the dynamic voxel spacing for training and testing purposes, we can calculate it based on the average length, denoted as L, of the bounding box in the x, y, and z directions.

$$VoxelSpacing = max(a, min(b, S_k \times L/L_k)) \qquad (1)$$

Here, S_k represents the median spacing of the training set (0.78), while a, b, and L_k are hyperparameters specifically chosen for this paper. In our study, these hyperparameters are set to 0.5, 1.4, and 110, respectively. Based on Eq. 1, it can be observed that in instances where the kidneys are exceptionally large, the model will be trained and inferred using a lower resolution. Conversely, when the kidneys are relatively small, the model will be trained and inferred using a higher resolution. The bounding box length L is calculated using the ground truth of kidney during training, and calculated using model' prediction during testing.

2.2   Model Architecture

The various stages of the model employ a similar model structure, all of which are variants of U-Net. The following Table 1 are the model parameters utilized for each stage:

2.3   Exclusion Learning

There are currently large amounts of publicly available datasets of abdominal CT scans, such as the Medical Segmentation Decathlon (MSD) dataset [1,7]. However, most of these datasets lack specific annotations for kidney or renal tumor segmentation and therefore cannot be directly applied to the task of this paper. In order to fully utilise the potential of these datasets, an exclusion learning method with pseudo-labels is employed. This approach is based on the principle that, in general, a pixel cannot belong to two different organs at the same time. For example, the MSD task03 dataset labels only the liver or liver tumor, and we can infer that regions belonging to the liver cannot be attributed to the kidney or kidney mass. Similarly, information from the MSD task09 (spleen), MSD task07 (pancreas) and MSD task 08 (liver vessel and tumors) datasets can also facilitate the segmentation of kidney tumors to some extent.

To start exclusion learning, a baseline model is trained. Then, this model is employed to conduct inference on the public dataset, comparing its predictions with the original non-kidney region labels provided by the dataset. A voxel is considered an over-segmentation voxel if it is labeled by both the model prediction and the original label of the MSD dataset. Cases exhibiting notable over-segmentation (with over $8000\,\mathrm{mm}^3$ over-segmented area) were selected, and the labels of the original non-renal regions from the public dataset were subtracted from the model's predictions, resulting in corrected pseudo-labels. Finally, a total of 29 CT images and corresponding pseduo labels are generated from MSD dataset were chosen for next round training.

2.4   Optimization

We employed nnUnet framework to train all models. DICE loss and CrossEntropy Loss is used as the loss function. Stochastic gradient descent (SGD) was used as an optimizer with a batch size of 2. All models are trained for 250k iterations with the initial learning rate of 1e−2, which takes around 2 days using a NVIDIA V100 GPU.

2.5 Pre-processing

We use nnUnet [3] auto-processing to process all the data, each case is clipped to the range calculated by nnUnet auto-planning. And then these data are subtracted from the mean and divided by the standard deviation.

2.6 Post-processing

For post-processing, we implemented a two-step approach. Firstly, we removed regions of the kidney where the predicted connectivity domain was less than $20,000\,\mathrm{mm}^3$. This step helped eliminate small, isolated regions that may not correspond to actual kidney structures. Next, similar to KiTS23 official post-processing method[2], we applied Gaussian filtering to the entire CT image, specifically targeting regions where the CT value was less than -30 HU. This filtering process effectively removed predicted values in regions with low CT attenuation, which are less likely to correspond to kidney tissues.

2.7 Dataset and Evaluation Metrics

We conducted validation of our method using the KiTS2023 competition data, consisting of a total of 489 renal enhancement CT images. Out of these, a randomly selected subset of 98 cases was used as the validation set. For evaluation purposes, we employed the Dice and Surface Dice metrics.

3 Experiments Results

3.1 Ablation Studies

Table 2 presents an analysis of the impact of each proposed component by comparing the DICE scores. The baseline model was constructed according to the details outlined in Sect. 2. We can observed that the dynamic resolution technique leads to improved performance, particularly in regions with high variability in scale and shape, such as tumor and cyst. Furthermore, the proposed exclusion learning architecture further enhances the performance, with significant improvements observed across all structures.

As shown in Fig. 4, our method shows greater robustness in some cases with specific shapes and scales.

3.2 Comparison of Existing Methods

We conducted a comparison with nnUnet-HighRes as a reference. The nnUnet-HighRes model utilizes parameters automatically generated by the nnUnet planner, where the voxel spacing is set to (1.0, 0.78, 0.78) and the patch size is set as (128, 128, 128). As shown in Table 3, proposed method has a significant improvement over nnUnet on all targets.

---

[2] https://github.com/neheller/kits23.

Table 2. Ablation results of dynamic resolution and exclusion learning. DR indicates the dynamic resolution, EL indicates the exclusion learning, TTA indicates the test-time augmentation. The inference time is the average results for 98 validation cases tested on single NVIDIA GeForce RTX 3080 GPU.

Method	Kidney+Tumor+Cyst	Tumor+Cyst	Tumor	Inference time
Baseline	0.9802	0.8437	0.7993	12 s/case
Baseline+DR	0.9810	0.8530	0.8077	15 s/case
Baseline+DR+EL	0.9831	0.8613	0.8188	15 s/case
Baseline+DR+EL+TTA	0.9833	0.8700	0.8290	73 s/case

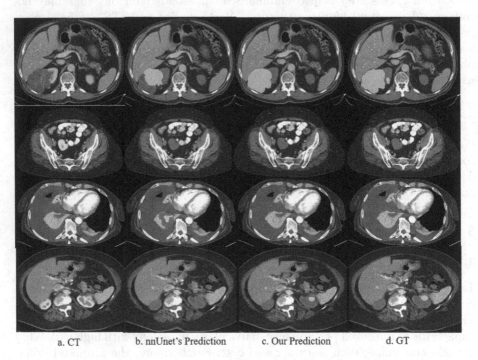

a. CT   b. nnUnet's Prediction   c. Our Prediction   d. GT

Fig. 4. Visualization of predictions generated by different approaches. From left to right are: a) the CT images. b) predictions by nnUnet. c) our predictions. d) the ground truth (GT).

Table 3. Proposed method vs. other methods. All methods do not use the test-time augmentation technique. The inference time is the average results for 98 validation cases tested on single NVIDIA GeForce RTX 3080 GPU.

Method	Kidney+Tumor+Cyst	Tumor+Cyst	Tumor	Inference time
nnUnet-HighRes [3]	0.9708	0.8200	0.7700	40 s/case
Ours	0.9831	0.8613	0.8188	15 s/case

4   Conclusion

In this study, we propose a new segmentation method for dealing with multi-scale kidney tumors. Compared to existing methods, the proposed method provides improvements in kidney, tumor prediction. It is expected to provide a powerful tool in large-scale radiomics analysis of renal tumor imaging and its relationship with tumor molecular features and disease-specific outcomes.

References

1. Antonelli, M., Reinke, A., Bakas, S., et al.: The medical segmentation decathlon. Nat. Commun. **13**(1), 4128 (2022). https://doi.org/10.1038/s41467-022-30695-9
2. Heller, N., Isensee, F., Maier-Hein, K.H., et al.: The state of the art in kidney and kidney tumor segmentation in contrast-enhanced CT imaging: results of the KiTS19 challenge. Med. Image Anal. **67**, 101821 (2021). https://doi.org/10.1016/j.media.2020.101821
3. Isensee, F., Jaeger, P.F., Kohl, S.A.A., et al.: nnU-Net: a self-configuring method for deep learning-based biomedical image segmentation. Nat. Methods **18**(2), 203–211 (2021). https://doi.org/10.1038/s41592-020-01008-z
4. Milletari, F., Navab, N., Ahmadi, S.A.: V-Net: fully convolutional neural networks for volumetric medical image segmentation. In: 2016 Fourth International Conference on 3D Vision (3DV), Stanford, CA, USA, pp. 565–571. IEEE (2016). https://doi.org/10.1109/3DV.2016.79
5. Ronneberger, O., Fischer, P., Brox, T.: U-Net: convolutional networks for biomedical image segmentation. In: Navab, N., Hornegger, J., Wells, W.M., Frangi, A.F. (eds.) MICCAI 2015. LNIP, vol. 9351, pp. 234–241. Springer, Cham (2015). https://doi.org/10.1007/978-3-319-24574-4_28
6. Shelhamer, E., Long, J., Darrell, T.: Fully convolutional networks for semantic segmentation. IEEE Trans. Pattern Anal. Mach. Intell. **39**(4), 640–651 (2017). https://doi.org/10.1109/TPAMI.2016.2572683
7. Simpson, A.L., Antonelli, M., Bakas, S., et al.: A large annotated medical image dataset for the development and evaluation of segmentation algorithms (2019). https://doi.org/10.48550/ARXIV.1902.09063
8. Zhao, Z., Chen, H., Wang, L.: A coarse-to-fine framework for the 2021 kidney and kidney tumor segmentation challenge. In: Heller, N., Isensee, F., Trofimova, D., Tejpaul, R., Papanikolopoulos, N., Weight, C. (eds.) KiTS 2021. LNCS, vol. 13168, pp. 53–58. Springer, Cham (2022). https://doi.org/10.1007/978-3-030-98385-7_8

Analyzing Domain Shift When Using Additional Data for the MICCAI KiTS23 Challenge

George Stoica[1,2(✉)], Mihaela Breaban[2], and Vlad Barbu[1]

[1] Sentic Lab, Iasi, Romania
george.stoica@senticlab.com
[2] Faculty of Computer Science, "Alexandru Ioan Cuza" University,
Iasi, Romania

Abstract. Using additional training data is known to improve the results, especially for medical image 3D segmentation where there is a lack of training material and the model needs to generalize well from few available data. Unlike transfer learning in which a model pretrained on huge datasets is fine-tuned for a specific task using limited data, we research the case in which we acquire supplementary training material and combine it with the original training data. However, the new data could have been obtained using other instruments and preprocessed such its distribution is significantly different from the target domain. Therefore, we study techniques which ameliorate domain shift during training so that the additional data becomes better usable for preprocessing and training together with the original data. We opt for using statistical criteria for reducing the distribution shift for domain adaptation in the context of having more data from the target domain than additional training data. Our results show that *transforming the additional data using histogram matching* has better results than using *simple normalization*. We achieved the 5th place on the official test dataset with a Dice score of 0.807 and Surface Dice of 0.691. On the validation set, we additionally report the Dice score for cysts (0.512) and kidney (0.946) besides the official metrics.

Keywords: 3D Segmentation · Domain Shift · Domain Adaptation

1 Introduction

The segmentation of renal structures (kidney, tumor, cyst) has gained interest in the recent years, starting from the KiTS19 Challenge [4] and continuing with KiTS21, KiPA22[1] and currently with KiTS23. The accurate segmentation of renal tumors and renal cysts is of important clinical significance and can benefit the clinicians in preoperative surgery planning.

---

[1] https://kipa22.grand-challenge.org/home/.

© The Author(s), under exclusive license to Springer Nature Switzerland AG 2024
N. Heller et al. (Eds.): KiTS 2023, LNCS 14540, pp. 22–29, 2024.
https://doi.org/10.1007/978-3-031-54806-2_4

Deep learning leverages on huge amount of training data for learning domain specific knowledge which can be used for predicting on previously unseen data. In medical image segmentation, a smaller amount of training data is available when compared to other domains of deep learning. Therefore, using additional training data has a greater impact on the end results.

In transfer learning, knowledge gathered from one or more source domains is used to execute new tasks in a different target domain. In deep learning, models are typically pretrained on a huge collection of data and fine-tuned on a downstream task. This procedure capitalizes on the previously learnt features to improve the performance on the new task, both in terms of results and computational resources needed for training. While pretrained models are available for many CV tasks on 2D images, for 3D medical images transfer learning is more challenging due to limited availability of publicly available annotated data. Using pretrained models on 2D slices from 3D volumes is certainly feasible, however with this approach the deep learning model learns only the intra-slice information and loses inter-slice correlations, resulting in weaker performance in general. Only recently [6] and [11] have studied pretraining for 3D medical images and achieved good results.

Domain adaptation is a particular case of transfer learning. In domain adaptation, the usual scenario entails learning from a source distribution and predicting on a related target distribution. The change in the distribution of the training dataset and the test dataset is called domain shift. In the case of supervised domain adaptation, labeled data from the target domain is available [10].

The medical image acquisition process is not uniform across different institutions and CT images may have different HU values and various amount of noise depending on the acquisition device, the acquisition time and other external factors. As a consequence, distribution shifts are easily encountered and this affects models that perform well on validation sets but encounter different data in practice. Creating a model which is robust to different types of distributions requires training on enough data, coming from all the target domains and learning domain invariant representations or a common shared space. Another approach is to preprocess and transform the production data, extracting features that match the already learnt distribution.

When training under domain shift and using two datasets with different distributions, the data ought to be preprocessed in order to mitigate the data mismatch error which happens due to the distribution shift. Characteristics of the target dataset have to be incorporated into the training dataset, which could be done either by collecting more data from the target distribution, or by artificial data synthesis. Our solution consists of transforming the additional data taken from the KiPA22 challenge to the target distribution which is represented by the data from the KiTS23 challenge.

This is a particular case of homogeneous domain adaptation in which the feature space between source and target are the same and labelled data from both domains is available. Unlike the usual scenario, in our case we have more labelled data from the target domain, 489 CTs and less labelled data from the source domain, 70 CTs. Therefore, our approach consists of translating the source distribution to the target distribution instead on focusing on learning a common shared space or domain invariant features.

We compare two transformations, dataset normalization, which preserves the original but different distribution, with histogram matching, which translates the additional data making both the source and target distribution the same.

2   Methods

Our approach consists of applying initial preprocessing to an additional dataset which was used for training. The aim of the preprocessing was to reduce the distribution shift between the additional data and the target domain, and will be fully-detailed in Sect. 2.2. After bridging the distributions of the original and additional data, we preprocess and normalize the whole data together and train a 3D U-Net [1] using multiple data augmentation techniques. Ultimately, we predict on the validation set and postprocess the prediction before evaluating them.

2.1   Training and Validation Data

Our submission uses the official KiTS23 training set, built upon the training and testing data from the KiTS19 [5] and KiTS21 competitions. In addition to the official KiTS23 data, our submission made use of the public KiPA22 competition training set [2,3,8,9].

The KiTS23 training dataset contains 489 CTs which include at least one kidney and tumor region and usually include both kidneys and optionally one or more cyst regions. In contrast, the KiPA22 training data contains only 70 CTs in which only the diseased kidney is selected. KiPA22 images have 4 segmentation targets: kidney, tumor, artery and vein. Unlike KiTS23, benign renal cysts are segmented as part of the kidney class for KiPA22. The initial preprocessing for the KiPA22 images consists of removing the artery and vein segmentation masks and keeping only the kidney and tumor class.

We have randomly chosen 342 images from KiTS23 and 70 images from KiPA22 for training and 147 images from KiTS23 for validation.

2.2   Preprocessing

Initial exploratory data analysis illustrate the fact that images from KiPA22 have a totally different distribution than images from the KiTS23 training set on the HU scale.

While the values for the KiTS23 CT images are mostly centered around -1000 and 0 on the HU scale, KiPA22 images are situated between 800 and 1500 while also having a visible different distribution (Fig. 1a). Training under domain shift using the original distribution for the second dataset is challenging, therefore we have taken steps towards ameliorating the effects of distribution shift.

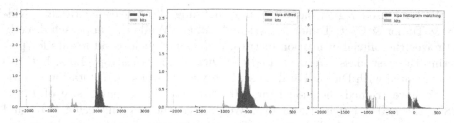

(a) Both datasets before preprocessing. (b) Shifting the mean and standard deviation. (c) Applying histogram matching.

Fig. 1. Histograms for both datasets before and after initial preprocessing.

To mitigate the impact of the huge distance between the values of the voxels, the simplest solution is shifting the mean and standard deviation of KiPA images to match those of KiTS (Fig. 1b). Nevertheless, the distributions are still visibly different, therefore we also applied histogram matching to transform the KiPA images to the KiTS domain (Fig. 1c).

We choose to transform the KiPA images because the test data will be made of images which are expected to match the original KiTS distribution. When training under domain shift, only data from the target distribution should be used for validation. Otherwise, we would risk having data mismatch error between the validation set and the test set.

For the KiPA dataset, shifting the mean maintains the original shape of the curve, scaled by the factor which changed the standard deviation, spreading the values evenly. Histogram matching, on the other side, is destructive and the HU values of voxels are spread unevenly to match the KiTS distribution. To choose the best transformation, we have created two datasets to evaluate them separately in order to choose the more suitable one.

1. **Dataset 1:** 342 images from KiTS and 70 images from KiPA whose values are shifted by changing the mean and standard deviation.
2. **Dataset 2:** the same 342 images from KiTS and 70 images from KiPA whose values are transformed by histogram matching.

For both datasets, the same preprocessing steps are applied, using the nnUNet framework [7]. Values are clipped at the 0.5th and 99.5th percentile to remove outliers. Then, images are normalized to have the mean 0 and standard deviation 1 and three order-interpolation is used to resample all images into a space of $0.7636 \times 0.7636 \times 0.7636 \, \text{mm}^3$.

2.3 Proposed Method

After preprocessing each dataset, we have trained the model using the default nnUNet v2 configuration for training, which uses a classic 3D U-Net. We have trained a single model on all the available data instead of opting for an ensemble trained on 5 folds.

We have used region based training, defining the 3 learning targets: Kidney & Tumor & Cyst, Tumor & Cyst and Tumor. While this approach directly optimizes the official evaluation metrics, it does not yield good results for predicting the cyst class. We have used the Dice & Cross Entropy Loss, but also experimented with Dice & Focal Loss which yielded worse segmentation results.

We have trained the model for 1000 epochs using a patch size of 128 × 128 × 128 and a batch size of two. We started the training using SGD and Nesterov momentum with a learning rate of 0.01 and used a Polynomial Learning Rate Scheduler to decrease the learning rate evenly until it reaches a value of 0.001 at the end of the 1000 epochs. To prevent overfitting, we applied multiple data augmentation techniques integrated in nnUNet: Rotation, Scaling, Gaussian noise, Gaussian blur, Random brightness, Gamma Correction and Mirroring.

For our custom postprocessing, we generate all the connected components from a prediction in order to determine all kidney candidates, and we choose two candidates using a heuristic for determining whether a connected component is a kidney or not. In our heuristic, we use the mean of all voxel positions to calculate the center of each connected component, and we choose the two candidates which have their center at a similar position on the y and z axis (sagital and axial plane) and are symmetric compared to the centre on the x axis (coronal plane). All other connected components are removed. Furthermore, we also generate all connected components of the lesion area and eliminate all areas with less than 10 voxels, considering that they are noise. Additionally, all voxels inside a lesion area are set to one class (tumor or cyst) by using majority voting, in order to remove all cases in which cyst and tumor voxels are predicted for the same lesion.

3   Results

We have trained on both **Dataset 1** and **Dataset 2** and have used 147 images from KiTS23 for validation. The results are displayed in Table 1. The official metrics used in competition are in *italic*, but we also report the Dice score for kidney and cyst segmentation.

Our experiments show that applying histogram equalization (**Dataset 2**) on the additional dataset improves the results for all the target metrics. Using simple normalization (**Dataset 1**) has better results only when calculating the Dice score for the kidney area. However, the Dice score for tumors and cysts is worse by 2 and 3%. Using the original distribution of the KiPA dataset results in a lower Dice score for cysts also because possible cysts are labeled as the kidney class. Nonetheless, since the two distributions are still very different even after normalization and preprocessing, the scores are heavily impacted.

Evaluating the results on both configurations, the model does not distinguish the cyst class and many cysts are classified as tumors. There is a class imbalance between cysts and tumors, as cysts generally encompass a smaller area. In our case, the low dice score for cysts is a result of many false positives. We presume that our learning target is the culprit, because we directly minimize the Dice and Cross Entropy loss for the whole segmentation area (kidney and

Table 1. Validation results for **Dataset 1** and **Dataset 2** and postprocessing results on **Dataset 2**. The test results on the official test dataset are also displayed for comparison.

Dataset/Dice score	kidney&masses	masses	kidney	tumor	cyst
Dataset 1	95.310904	79.072143	**94.101086**	76.891783	17.012944
Dataset 2	**95.453839**	**80.760511**	94.024749	**78.960431**	**20.766421**
Postprocessing	**97.075388**	**85.229431**	94.679921	82.734208	51.224890
Test results	94.7	76	–	71.3	–

masses), the lesion area (masses, both tumor and cyst) and ultimately, tumor. As a consequence, the cyst area is indirectly learnt, therefore the accuracy is lower. To make the model discriminate between the two classes and reduce the false positives, we suggest changing the learning target to directly minimize the Dice and Cross Entropy loss for the cyst class.

By applying postprocessing on **Dataset 2** we greatly increase all our results, especially the Dice score for the cyst areas which is more than doubled. This is due to the fact that we eliminate spurious voxels with the cyst class, predicted by our model in tumor areas, or tumor voxels predicted in cyst areas. The Dice score for tumors is also increased by 3.7 for the same reasons as above, but the impact of postprocessing is lower. Nevertheless, all official metrics benefit from postprocessing and provide a promising growth.

We have achieved the 5th place on the official testing dataset with a Dice score of 0.807 and Surface Dice of 0.691. The Dice score is the average of the three official metrics displayed on the last row of Table 1 and the results for the kidney and cyst area are not available on the official leaderboard. Compared to the validation results, there is a decrease in score, especially for the masses and tumor areas. Motivated by the difference in Dice score for masses and tumor and considering that the cyst class is usually less predominant and has a smaller volume in general, we presume that the worse performance is explained by less accurate predictions for the tumor area on the provided test set. The culprit is most probably a harder test dataset, with cases for which tumors are not easily identifiable. Nevertheless, this illustrates that further research and model fine-tuning is to be done to improve the performance and robustness of the model.

For training and inference we have used a workstation with an RTX 3090 GPU, an AMD Ryzen Threadripper 2970WX 24-Core Processor CPU, SSD and 31 GB RAM memory available. Training the model took around 3.4 days. For prediction, the average inference time was less than 10 min per case.

4   Discussion and Conclusion

Distribution shift is a recurring issue that often appears in practice. The model underperforms when encountering data coming from a different distribution other than the source distribution on which the model was trained. The best performance is achieved when the source and target distribution is the same, but this is not always the case in real life scenarios. Therefore, we propose preprocessing the source distribution using statistical criteria in the context of 3D medical images with similar features but different values due to different acquisition processes.

We have explored suitable transformation techniques for mitigating distribution shift when using additional data for the kidney tumor 3D segmentation task. We have identified histogram matching as an initial preprocessing step of artificial data synthesis that completely transforms the source distribution to the target domain. Compared to simple normalization, this approach has the advantage of training only on the target distribution, which improves the results, especially for the least frequent classes, cyst and tumor. Our postprocessing techniques also greatly increase the score by correcting visible mistakes done by our model in a deterministic manner using domain specific knowledge.

Our approach yields good results on the official test dataset, achieving the 5th place on the final Leaderboard. We believe that more stable results can be achieved by training an ensemble, and the discriminative power between cysts and tumors can be enhanced by changing the training target and using techniques that deal with class imbalance.

References

1. Çiçek, Ö., Abdulkadir, A., Lienkamp, S.S., Brox, T., Ronneberger, O.: 3D U-Net: learning dense volumetric segmentation from sparse annotation. In: Ourselin, S., Joskowicz, L., Sabuncu, M., Unal, G., Wells, W. (eds.) MICCAI 2016. LNCS, pp. 424–432. Springer, Cham (2016). https://doi.org/10.1007/978-3-319-46723-8_49
2. He, Y., et al.: Dense biased networks with deep priori anatomy and hard region adaptation: semi-supervised learning for fine renal artery segmentation. Med. Image Anal. **63**, 101722 (2020)
3. He, Y., et al.: Meta grayscale adaptive network for 3D integrated renal structures segmentation. Med. Image Anal. **71**, 102055 (2021)
4. Heller, N., et al.: The state of the art in kidney and kidney tumor segmentation in contrast-enhanced CT imaging: results of the KiTS19 challenge. Med. Image Anal. **67**, 101821 (2021)
5. Heller, N., et al.: The KiTS19 challenge data: 300 kidney tumor cases with clinical context, CT semantic segmentations, and surgical outcomes. arXiv preprint arXiv:1904.00445 (2019)
6. Huang, Z., et al.: STU-Net: scalable and transferable medical image segmentation models empowered by large-scale supervised pre-training. arXiv preprint arXiv:2304.06716 (2023)
7. Isensee, F., Jaeger, P.F., Kohl, S.A., Petersen, J., Maier-Hein, K.H.: nnU-Net: a self-configuring method for deep learning-based biomedical image segmentation. Nat. Methods **18**(2), 203–211 (2021)
8. Shao, P., et al.: Laparoscopic partial nephrectomy with segmental renal artery clamping: technique and clinical outcomes. Eur. Urol. **59**(5), 849–855 (2011)
9. Shao, P., et al.: Precise segmental renal artery clamping under the guidance of dual-source computed tomography angiography during laparoscopic partial nephrectomy. Eur. Urol. **62**(6), 1001–1008 (2012)

10. Wang, M., Deng, W.: Deep visual domain adaptation: a survey. Neurocomputing **312**, 135–153 (2018)
11. Wang, Y., et al.: SwinMM: masked multi-view with swin transformers for 3D medical image segmentation. arXiv preprint arXiv:2307.12591 (2023)

A Hybrid Network Based on nnU-Net and Swin Transformer for Kidney Tumor Segmentation

Lifei Qian, Ling Luo, Yuanhong Zhong, and Daidi Zhong[✉]

Chongqing University, Chongqing, China
daidi.zhong@cqu.edu.cn

Abstract. Kidney cancer is one of the most common cancers. Precise delineation and localization of the lesion area play a crucial role in the diagnosis and treatment of kidney cancer. Deep learning-based automatic medical image segmentation can help to confirm the diagnosis. The traditional 3D nnU-net based on convolutional layers is widely used in medical image segmentation. However, the fixed receptive field of convolutional neural networks introduces an induction bias limiting their ability to capture long-range spatial information in input images. The Swin Transformer addresses this limitation by leveraging the global contextual modeling ability obtained through self-attention computation. However, it requires a large amount of training data and lacks in local feature encoding. To overcome these limitations, our paper proposes a hybrid network structure called STransUnet, which combines the nnU-net with Swin Transformer. STransUnet retains the local feature encoding capability of nnU-net while introducing the Swin Transformer to capture a broader range of global contextual information, resulting in a more powerful modeling ability for image segmentation tasks. In the KiTS23 challenge, our average Dice and average Surface Dice of segmentation on the test are 0.801 and 0.680 ranked the 6th and 8th respectively and our Tumor Dice is 0.687.

Keywords: Convolutional neural network · nnU-Net · Swin Transformer

1 Introduction

Kidney cancer is one of the most common cancers, with over 430,000 people diagnosed each year, of which approximately 180,000 cases result in death [5]. Computerized tomography (CT) imaging plays a crucial role in the diagnosis and understanding of the characteristics of kidney tumors. The segmentation of kidney tumors is an important basis for doctors to determine diagnosis and treatment plans. The delineation and segmentation of most medical images are performed by radiologists. However, due to human subjectivity, significant variations among doctors, and factors such as fatigue, the accuracy and quantity of

© The Author(s), under exclusive license to Springer Nature Switzerland AG 2024
N. Heller et al. (Eds.): KiTS 2023, LNCS 14540, pp. 30–39, 2024.
https://doi.org/10.1007/978-3-031-54806-2_5

image annotations by humans are limited. This greatly restricts the ability of the medical field to progress towards more digitized and personalized healthcare. Deep learning-based medical image segmentation, through learning and training on a large number of parameters, can achieve automatic segmentation of medical images and has many applications in clinical quantification, treatment, and surgical planning [18].

Convolutional Neural Networks (CNNs) trained on large annotated datasets have shown superior performance, surpassing traditional algorithms and even human capabilities. CNNs have many outperforming representations such as AlexNet [9], VGGNet [13], Inception Net [15] and ResNet [7]. Among all CNNs, nnU-Net [8] has achieved remarkable results in the segmentation of three-dimensional medical images. The key factor behind the success of nnU-Net is their inductive bias in dealing with scale-invariant local visual structures. While this inherent locality (limited receptive field) brings efficiency to nnU-Net, it weakens their ability to capture long-range spatial information in input images, thereby bottlenecking their performance [11]. This calls for an alternative architectural design that can model relationships between distant pixels for better representation learning.

To overcome the limitations of fixed receptive fields, one approach is to integrate attention mechanisms into CNN-inspired architectures [2,4]. These attention-based models have become an attractive solution as they can encode long-range dependencies and learn efficient feature representations. The global context modeling ability of Transformer [16] is crucial for accurate medical image segmentation since it allows for effective segmentation of regions distributed across a large field by capturing relationships between distant pixels. By operating on a set of image patches, Alexey Dosovitskiy et al. proposed the Vision Transformer (ViT) [3], which completely replaces standard convolutions in deep neural networks. However, the Transformer architecture was originally proposed in the field of natural language processing (NLP), where the pixel resolution of images is much higher than the resolution of text paragraphs. When used for visual tasks, dense predictions at the pixel level are required, which is challenging for Transformer on high-resolution images due to the quadratic complexity of self-attention computations with respect to image size. In order to control the model size and inference time, our paper adopts the Swin Transformer proposed by Liu et al. [10] to extract contextual information. Swin Transformer is an image processing model based on the Transformer architecture, specifically designed for handling high-resolution image tasks. It improves upon the Vision Transformer to address the computational and memory overhead of traditional Transformers when dealing with large-sized images. It divides the image into windows and performs self-attention operations within each window. This approach reduces the length of the input sequence, thereby reducing computational and memory costs while being able to handle high-resolution images. Meanwhile, Transformers face challenges in handling local information and extracting local features, where traditional approaches like nnU-Net may be more suitable. Therefore, the application of Transformer in image segmentation is often accompanied by convolutional layers, such as TransUNet proposed by Chen et al. [1], Transformer-Unet by Sha et al. [12], and Swin UNETR by Hatamizadeh et al. [6].

To encode contextual information without sacrificing local feature extraction and to ensure a manageable number of parameters without compromising accuracy, we introduces STransUnet.

2   Methods

Our model adopts a hybrid approach, combining the nnU-Net framework and the Swin Transformer framework. It adopts an encoder-decoder architecture, where the intermediate output layers of the encoder are composed with the outputs of nnU-Net and Swin Transformer. The decoder follows the architecture of nnU-Net, and there are skip connections between the encoder and decoder. The network architecture is illustrated in the diagram below (Fig. 2):

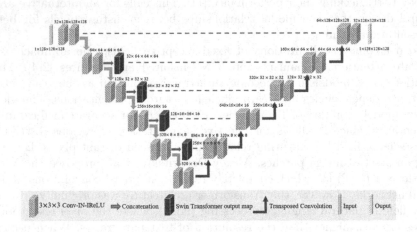

Fig. 1. STransUnet architecture

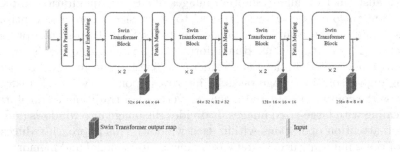

Fig. 2. Architecture of the Swin Transformer Component

2.1 Training and Validation Data

Our submission made use of the official KiTS23 training set alone. We randomly selected 23 cases from the official KiTS23 training set as the test set, 372 cases as the training set, and 94 cases as the validation set.

2.2 Preprocessing

The original image dimensions range from (512, 512, 1059) to (512, 512, 29). The voxel sampling spacing varies for each axis, with the x-axis ranging from 0.39 to 1.04, the y-axis ranging from 0.39 to 1.04, and the z-axis ranging from 0.50 to 5.00. The image dimensions determine the size of the image, which in turn affects the computational complexity and inference time. Different sampling spacings affect the resolution of the image. A larger sampling spacing results in lower resolution and less detailed information but retains more contextual information. A smaller sampling spacing provides higher resolution and more detailed information but sacrifices contextual information. Due to the non-uniform voxel sampling spacing in the dataset, we followed the approach of nnU-Net by selecting the median value of the sampling spacing for each axis as the target sampling spacing. In this paper, the final target sampling spacing is selected as (1.00, 0.78, 0.78). Since we adopted the low-resolution nnU-Net, the voxel target sampling spacing is chosen as (2.38, 1.86, 1.86). According to the selected target sampling spacing, third-order spline interpolation is employed for resampling.

Considering the limited memory budget of the GPU, the non-zero region cropping technique can be applied to reduce memory consumption by focusing on the region of interest. Additionally, to reduce memory requirements and computational complexity due to the large size of the images, the image can be divided into patches, decomposing the problem into handling multiple smaller patches. In this paper, the patch size is (128, 128, 128), and the batch size is set to 2.

Furthermore, intensity normalization is performed. Since the intensity scale of CT scans is absolute, all CT images are automatically normalized based on the statistical data of the entire dataset. We collected all intensity values present in the training dataset and standardizes the entire dataset by clipping these intensity values at the 0.5th percentile and 99.5th percentile, followed by z-score normalization based on the mean and standard deviation of all collected intensity values. The implementation of z-score normalization is as follows:

$$Z = \frac{(X - \mu)}{\sigma} \tag{1}$$

where X is the original data, μ is the mean, and σ is the standard deviation.

To train large neural networks with limited training data, preventing overfitting is crucial. To augment the training set and prevent overfitting, the following data augmentation techniques are applied in this paper: random rotation, random scaling, random elastic deformation, gamma correction, and mirroring. Random rotation involves randomly rotating the image by a certain angle, which increases the model's ability to recognize objects from different angles. By rotating the image, samples with different angles and orientations can be generated. In this study, the probability of random rotation is set to 0.2, with a rotation angle range of $(-0.52, 0.52)$.

Random scaling involves randomly scaling the image. By varying the size and scale of the image, the model's ability to segment objects of different scales is enhanced. The probability of random scaling is set to 0.2, with a scaling range of $(0.7, 1.4)$.

Random elastic deformation involves randomly deforming the image. By introducing random deformation parameters within a certain range, the model's tolerance to deformations and distortions is increased, improving its robustness. This method simulates the influence of various deformation factors on the image in the real world, such as deformation, distortion, shape distortion, etc. In this study, the probability of random elastic deformation is set to 0.2, with a deformation strength parameter range of $(0, 900)$ and a smoothness parameter range of $(9.0, 13.0)$.

Gamma correction is an enhancement technique used to adjust the brightness and contrast of the image. By changing the pixel value distribution of the image, the model's adaptability to different brightness conditions is improved. The implementation formula is as follows:

$$Y = cX^{\gamma} \tag{2}$$

where Y is the transformed output, X is the pixel value of the image, c is the grayscale scaling coefficient (set to 1 in this study), and γ is the adjustment constant that controls the degree of scaling for the gamma transformation. It has a significant impact on the characteristics of the transformation function. In this study, γ ranges from 0.7 to 1.5, and the probability of gamma correction is 0.3. When $\gamma > 1$, grayscale compression is applied to brighter images, while when $\gamma < 1$, contrast enhancement is applied to darker images, strengthening the image details.

Mirroring involves flipping the image horizontally or vertically, generating mirror-symmetric samples. Mirroring increases the invariance and robustness of the model to object segmentation.

2.3 Proposed Method

We selected the first stage of the low-resolution cascaded nnU-Net as the baseline network. Swin Transformer is integrated into the nnU-Net framework to incorporate long-range feature information to overcome the limitations of nnU-Net's local receptive field. In order to combine the feature maps extracted by nnU-Net and Swin Transformer, while maintaining the integrity of the nnU-Net framework, we concatenated the feature maps of second, third, fourth and fifth levels from nnU-Net and first, second, third and fourth levels from Swin Transformer along the channel dimension. The final output feature map sizes of the encoder are as follows (Table 1):

Table 1. Size of the fused encoder output feature maps

Output Level	Size
1	(32, 128, 128, 128)
2	(96, 64, 64, 64)
3	(192, 32, 32, 32)
4	(384, 16, 16, 16)
5	(576, 8, 8, 8)

The overall network framework is illustrated Fig. 1. Our network architecture adopts an encoder-decoder structure, which is a common neural network architecture consisting of two parts: an encoder and a decoder. The encoder encodes the input image into a fixed-length vector representation, while the decoder uses this vector representation to generate the output image. In this study, training is performed in an end-to-end manner, where the entire process from input to output is completed by a single model without any apparent stages or manual intervention. The input image is directly mapped to the output, eliminating the tedious process of manually designing features or intermediate steps, making the entire training process more concise and efficient.

In the encoder stage, two stacked convolutional layers are used to encode the feature maps. The first convolutional layer has a stride of 2 and an output channel twice that of the input channel, which performs downsampling. The second convolutional layer has a stride of 1 and an output channel equal to the input channel, which further extracts features. The convolutional layers have a kernel size of 3, and each is followed by an Instance Normalization layer and a Leaky ReLU layer to normalize the feature sequence and increase the model's nonlinearity. The construction of the Swin Transformer part remains the same as the original Swin Transformer encoder part. Swin Transformer Blocks are composed of multi-head self-attention modules with regular and shifted windowing configurations. After the Swin Transformer Block, the output features maps of each stage are concatenated with the corresponding feature maps of nnU-Net.

In the decoder stage, upsampling is first performed using a transposed convolution to reduce the channel and double the resolution. The transposed convolution has a kernel size and stride of 2. Then, a decoding block is formed by stacked two convolutional layers, where the output channel of the first convolutional layer is halved and the output channel of the second convolutional layer remains the same. Both layers have a stride of 1 and are followed by an Instance Normalization layer and a Leaky ReLU layer.

StransUnet employs a deep supervision strategy, which is a training technique aimed at improving the training effectiveness and optimizing gradient propagation in neural networks. In addition to the final segmentation result, additional supervision signals are added to the intermediate layers of the neural network to better guide network learning. The loss function is computed on the four output layers, and the predicted feature maps from these outputs, obtained through a

$1 \times 1 \times 1$ convolution and a Softmax layer, are compared with the ground truth labels. Deep supervision can alleviate the problem of vanishing or exploding gradients, allowing gradients to propagate better in the network. Moreover, since deep supervision can guide the learning process more directly at different levels, it can help the network converge faster. The four-level losses in this study are weighted to obtain the final loss, with weights assigned based on the output sizes. Specifically, the loss weights from level 1st to level 4th are 0.53, 0.27, 0.13, and 0.07.

In this paper,we chose Cross-entropy loss [17] and Dice loss [14] as our composite original loss and introduced weight decay, the final expression is as follows:

$$L = -\frac{1}{N} \sum_{i=1}^{N} (w_{ce} \cdot y_i \cdot \log(\hat{y_i}) + w_{dice} \cdot \frac{2 \cdot y_i \cdot \hat{y_i}}{y_i + \hat{y_i}}) + \lambda \cdot ||w||^2 \qquad (3)$$

where L is the loss function with weight decay, N is the number of samples in the batch, y_i represents the ground truth segmentation mask for sample i, $\hat{y_i}$ represents the predicted segmentation mask for sample i, w_{ce} and w_{dice} are the weighting factors for the Cross-Entropy loss and Dice loss, respectively. The Cross-Entropy loss term measures the pixel-wise dissimilarity between the predicted and ground truth masks, while the Dice loss term calculates the overlap between them. The weights control the relative importance of each loss term in the overall loss function (in this paper, both set to 1). λ is the coefficient of the weight decay term controlling the strength of regularization (set to $3\mathrm{e}-05$ in this paper), and $||w||^2$ is the squared L2 norm of the model's weight parameters, representing the sum of squared weights. By adding the weight decay term to the original loss function, the optimization process not only minimizes the original loss but also constrains the size of the weight parameters through the influence of the regularization term. This way, the optimization process tends to select smaller weight values, limiting the complexity of the model and reducing the risk of overfitting.

The optimization algorithm used in this paper is Stochastic Gradient Descent (SGD) with Nesterov momentum. SGD estimates the gradient of the entire training set by using random subsets of training samples. For each mini-batch, the gradients of the model parameters are computed for the loss function. The gradient represents the rate and direction of change of the loss function at the current parameter values. SGD with momentum accelerates convergence and reduces oscillation by introducing the concept of momentum, which can be understood as the inertia of parameter updates, similar to the concept of momentum in physics. Nesterov momentum is an improvement over standard momentum, used to estimate the gradient more accurately and guide parameter updates better. Unlike standard momentum, Nesterov momentum first calculates the gradient of the position obtained by adding the momentum term to the current parameter position, and then uses that gradient for parameter updates. This additional step improves the accuracy of gradient estimation and thereby improves the direction of parameter updates. Nesterov momentum converges more quickly to the optimal solution, especially in cases with large curvature or flat regions. In this paper, the momentum is set to 0.99 and the initial learning rate chosen is 0.01. In all experiments, the official evaluation code is used, which computes the average scores of Dice and Surface Dice metrics. After the model training is completed, inference is performed on the test set, and the model with the highest average scores for Dice and Surface Dice metrics is selected as the final submission model.

3 Results

For simplicity, in the table of this paper, we represent the Dice scores of each Hierarchical Evaluation Class (HEC) for "Kidney + Tumor + Cyst", "Tumor + Cyst", and "Tumor Only" as D1, D2, D3, and their average as MD. Similarly, the Surface Dice metric scores are represented as S1, S2, S3, and their average as MS. The experimental results are shown in the following Table 2:

Table 2. Results of nnU-Net and StransUnet

Network	D1	D2	D3	MD	S1	S2	S3	MS
nnU-Net	**0.9751**	0.8787	0.8484	0.9001	**0.9467**	0.7651	0.7227	0.8115
StransUnet	0.9748	**0.8822**	**0.8541**	**0.9037**	0.9437	**0.7685**	**0.7310**	**0.8144**

We ran 1000 epochs using a 4090 GPU. For nnU-net, the time taken for one epoch was approximately 280 s, and the inference time for one case was around 27 s. For StransUnet, the time taken for one epoch was approximately 380 s, and the inference time for one case was around 30 s.

On the official test set of KiTS23, the prediction results of StransUnet are as follows (Table 3):

Table 3. Results on official test set

Network	MD	MS	D3
StransUnet	0.801	0.680	0.687

Compared to the test set we partitioned ourselves, the official results have shown a certain degree of decline, indicating that our model's generalization ability needs improvement, and we need to expand the training dataset. Since our current training set consists of 372 cases, accounting for 0.76 of all data, we will consider increasing the proportion of the training set and increasing the probabilities of data augmentation in the future.

4   Discussion and Conclusion

From the experiment results, it can be observed that our model performs better than nnU-Net in the two HECs with smaller segmentation labels, "Tumor + Cyst" and "Tumor Only". This indicates that incorporating the output feature maps of Swin Transformer during the feature extraction process in the encoder is beneficial for segmenting small target regions. Swin Transformer performs attention calculation in each shifted window. In this paper, the window size of Swin Transformer is set to $(7, 7, 7)$, which can be understood as the receptive field size of Swin Transformer being $(7, 7, 7)$. By integrating contextual information through sliding windows, the model retains the detailed and complete nature of local features to some extent, thereby improving the model's performance.

However, our official results are lower than the experimental results, showing that our model's generalization ability and robustness need further improvement. The first measure is to adjust the proportions of the training set, validation set, and test set in the future. The second measure is to increase the probabilities of data augmentation transformations to expand the dataset. Finally, we are considering that the complexity of the model may be too high. In the future, we will reduce the number of stacked convolutional layers to decrease the model's parameter count and improve its robustness.

Our model is trained in an end-to-end manner and is a non-cascaded model that does not require manual intervention. In the future, we plan to perform model reparameterization. During the inference stage, we aim to simplify operations such as normalized convolution, continuous convolution and convolution concatenation into a single convolution using the principle of reparameterization. This will reduce the inference time of the model and make our model more concise.

Acknowledgment. This work is supported by the National Key Research and Development Program of China (Grant no: 2021YFC2009200).

References

1. Chen, J., et al.: TransUNet: transformers make strong encoders for medical image segmentation. arXiv preprint arXiv:2102.04306 (2021)
2. Devlin, J., Chang, M.W., Lee, K., Toutanova, K.: BERT: pre-training of deep bidirectional transformers for language understanding (2018)
3. Dosovitskiy, A., Beyer, L., Kolesnikov, A., Weissenborn, D., Houlsby, N.: An image is worth 16×16 words: transformers for image recognition at scale (2020)
4. Fedus, W., Zoph, B., Shazeer, N.: Switch transformers: scaling to trillion parameter models with simple and efficient sparsity (2021)
5. Bray, F., Ferlay, J., Soerjomataram, I., Siegel, R.L.: Global cancer statistics 2018: globocan estimates of incidence and mortality worldwide for 36 cancers in 185 countries. CA: Cancer J. Clin. (2018)
6. Hatamizadeh, A., Nath, V., Tang, Y., Yang, D., Roth, H.R., Xu, D.: Swin UNETR: swin transformers for semantic segmentation of brain tumors in MRI images. In: Crimi, A., Bakas, S. (eds.) BrainLes 2021. LNCS, vol. 12962, pp. 272–284. Springer, Cham (2021). https://doi.org/10.1007/978-3-031-08999-2_22

7. He, K., Zhang, X., Ren, S., Sun, J.: Deep residual learning for image recognition. In: IEEE Conference on Computer Vision and Pattern Recognition (2016)
8. Isensee, F., Jaeger, P.F., Kohl, S.A.A., Petersen, J., Maier-Hein, K.H.: nnU-net: a self-configuring method for deep learning-based biomedical image segmentation. Nat. Methods (2021)
9. Krizhevsky, A., Sutskever, I., Hinton, G.: ImageNet classification with deep convolutional neural networks. In: Advances in Neural Information Processing Systems, vol. 25, no. 2 (2012)
10. Liu, Z., et al.: Swin transformer: hierarchical vision transformer using shifted windows (2021)
11. Matsoukas, C., Haslum, J.F., Sderberg, M., Smith, K.: Is it time to replace CNNs with transformers for medical images? (2021)
12. Sha, Y., Zhang, Y., Ji, X., Hu, L.: Transformer-UNet: raw image processing with unet. arXiv preprint arXiv:2109.08417 (2021)
13. Simonyan, K., Zisserman, A.: Very deep convolutional networks for large-scale image recognition. Comput. Sci. (2014)
14. Sudre, C.H., Li, W., Vercauteren, T., Ourselin, S., Jorge Cardoso, M.: Generalised dice overlap as a deep learning loss function for highly unbalanced segmentations. In: Cardoso, M.J., et al. (eds.) DLMIA/ML-CDS -2017. LNCS, vol. 10553, pp. 240–248. Springer, Cham (2017). https://doi.org/10.1007/978-3-319-67558-9_28
15. Szegedy, C., Liu, W., Jia, Y., Sermanet, P., Rabinovich, A.: Going deeper with convolutions. In: 2015 IEEE Conference on Computer Vision and Pattern Recognition (CVPR) (2015)
16. Vaswani, A., et al.: Attention is all you need. arXiv (2017)
17. Yi-de, M., Qing, L., Zhi-Bai, Q.: Automated image segmentation using improved PCNN model based on cross-entropy. In: Proceedings of 2004 International Symposium on Intelligent Multimedia, Video and Speech Processing, pp. 743–746. IEEE (2004)
18. Zhou, S.K., et al.: A review of deep learning in medical imaging: image traits, technology trends, case studies with progress highlights, and future promises (2020)

Leveraging Uncertainty Estimation for Segmentation of Kidney, Kidney Tumor and Kidney Cysts

Zohaib Salahuddin[1]([✉]), Sheng Kuang[1], Philippe Lambin[1,2],
and Henry C. Woodruff[1,2]

[1] The D-Lab, Department of Precision Medicine, GROW-School for Oncology
and Reproduction, Maastricht University, Maastricht, The Netherlands
`z.salahuddin@maastrichtuniversity.nl`
[2] Department of Radiology and Nuclear Medicine, GROW-School for Oncology,
Maastricht University Medical Center, Maastricht, Netherlands

Abstract. In the field of medical imaging, computed tomography (CT) scans have become crucial for the detection and management of anatomical abnormalities. This study presents an improved cascaded nnUNet framework incorporating a cropping strategy and uncertainty estimation for effective segmentation of kidneys, kidney tumors, and kidney cysts in computed tomography scans. The proposed method is evaluated on the KiTS23 dataset, consisting of 489 CT scans with accompanying masks for the kidney, tumor, and cyst. We exploited a low-resolution nnUNet for initial kidney segmentation, and the resulting predictions were used to crop a bounding box area to decrease data dimensionality, which facilitated faster training and inference. A cyclic learning rate was applied along with posterior sampling of the weight space, enabling an ensemble of five models from different training cycles. This approach showed superior performance, particularly in the segmentation of tumors and masses, as compared to other models such as the standard nnUNet, the cascaded nnUNet, and the BANet. Moreover, our ensemble model, including models from different training cycles, indicated a strong correlation between predicted uncertainty maps and false positive detection, holding promising potential for enhanced clinical utility.

Keywords: 3D UNet · Uncertainty Estimation · Multi-stage Segmentation · Kidney Tumor · Interpretability

1 Introduction

In recent decades, with the advancement of medical imaging techniques, computed tomography (CT) scans have started to play a major role in clinical settings for the detection of anatomical abnormalities, as well as for the management and follow-up of patients, especially in medical oncology. The number of

Z. Salahuddin and S. Kuang—These authors contributed equally as first authors.

© The Author(s), under exclusive license to Springer Nature Switzerland AG 2024
N. Heller et al. (Eds.): KiTS 2023, LNCS 14540, pp. 40–46, 2024.
https://doi.org/10.1007/978-3-031-54806-2_6

patients diagnosed with kidney cancer is increasing with more than 430,000 patients diagnosed each year [1]. Active surveillance of potentially malignant tumors using medical imaging also results in increased radiological burden. It is not possible to determine the tumor characterization as malignant or benign directly from the CT scan with the use of automated segmentation methods [8]. There is a need for an automated segmentation algorithm that is able to accurately detect and segment the kidney and the relevant tumors and cysts. The automated delineations of these regions of interest also pave the way for radiomics studies for characterizing tumors and cysts as benign or malignant [7].

Deep Neural Networks have demonstrated state-of-the-art performance in various segmentation tasks for medical image analysis [9]. No-new-UNet (nnUNet) is as a self-configuring method that makes automatic design choices related to pre-processing, network architecture, and hyper-parameter tuning [5]. Different variants of nnUNet have demonstrated state-of-the-art performance on previous editions of the Kidney and Kidney tumor segmentation challenge (KiTS) [2,3]. A modified cascaded coarse-to-fine framework of nnUNet demonstrated the best performances on the 2021 editions of the KiTS challenge. KiTS 2023 edition uses an expanded training set comprising 489 cases and a test set comprising 110 cases. KiTS 2023 dataset contains CT scans in both nephrogenic contrast and late arterial phases.

The is a lot of heterogeneity present in the medical data and the deep learning-based segmentation algorithms can fail silently when they encounter out-of-distribution data thereby undermining the reliability of these algorithms [6]. It is important to incorporate uncertainty estimation in the segmentation algorithms to avoid silent failures and improve robustness and reliability for clinical adoption. There is also a positive correlation between uncertainty and false positives [10].

In this paper, we propose a modified version of a cascaded nnUNet framework that incorporates cropping to reduce the dataset and uncertainty estimation for the segmentation of kidney, kidney tumors, and kidney cysts in computed tomography scans.

2 Material and Methods

2.1 Data

We utilized the KiTS23 dataset which is comprised of 489 CT scans, obtained from patients suspected of renal malignancy, collected between 2010 and 2022 at an M Health Fairview medical center. Each CT scan in this dataset is accompanied by the corresponding masks for the kidney, tumor, and cyst. Each CT scan includes between 29 to 1059 axial slices, with a voxel size ranging from $(0.61 \times 0.61 \times 0.5 \text{ mm}^3)$ to $(1.0 \times 1.0 \times 5.0 \text{ mm}^3)$ (Fig. 1).

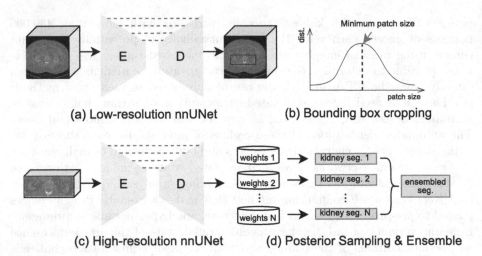

(a) Low-resolution nnUNet

(b) Bounding box cropping

(c) High-resolution nnUNet

(d) Posterior Sampling & Ensemble

Fig. 1. Our proposed method for kidney segmentation. (a) Low-resolution uuUNet for first-stage kidney segmentation. (b) Bounding box generation based on minimum patch distribution (c) High-resolution nnUNnet for second-stage multi-class segmentation (d) Ensemble prediction after posterior sampling of network weights.

2.2 Proposed Method

The CT images are resampled to the median resolution of 0.78 mm $\times$ 0.78 mm $\times$ 1 mm^3. The images are resampled using spline interpolation, while the corresponding segmentations are resampled using nearest-neighbor interpolation. The intensity values of the CT images are clipped at the 0.5 and 99.5 percentiles, and z-score normalization is applied.

In the first step, the low-resolution nnUNet is utilized, and five-fold cross-validation is performed on the training set. The low-resolution nnUNet is trained for 400 epochs. The predictions from nnUNet are then used to crop a bounding box area that includes the kidney structures from both kidneys. It is ensured that the minimum bounding box area is equivalent to at least the median area of the bounding boxes to guarantee sufficient coverage of the kidney structures, even if the low-resolution nnUNet fails. This cropping step reduces the dimensionality of the data and enables faster training and inference.

In the second step, the dataset was divided into a training set with 391 subjects and a validation set with 98 subjects. The cropped full resolution nnUNet was trained for 1000 epochs. We also evaluated the performance of boundary-aware network (BANet) [4]. We implemented a cyclic learning rate strategy and sampled posterior model weights around multiple local peaks for training our model [10]. We allocated a total training budget of 1200 units to ensure adequate model convergence within each training iteration. We segmented this training budget into three distinct training cycles. To estimate the uncertainty of our model's predictions, we saved 10 model checkpoints per training cycle, resulting in a total of 30 checkpoints. From these 30, we selected the top five best-performing checkpoints across all cycles, ensuring representation from each of the three training cycles. To derive the final prediction, we took the average of the probability outputs from these five checkpoints. For the post-processing stage, we retained the two most substantial connected components result.

The investigation into the relationship between predicted uncertainty and false positives was conducted for tumors and masses (comprising both cysts and tumors). The predicted segmentation masks were multiplied by the uncertainty predictions. Uncertainty density was determined by summing the uncertainty and then dividing by the volume. A threshold, determined by percentile, was applied to examine the relationship between the true positive rate of both retained and excluded lesions.

Table 1. Dice similarity metric (DSC) and Surface Dice (SD) for different models on the validation dataset.

Model	DSC Kidney	DSC Masses	DSC Tumor	SD Kidney	SD Masses	SD Tumor
Low Resolution nnUNet	0.971	0.843	0.781	0.942	0.740	0.681
Cascade nnUNet	**0.979**	0.854	0.802	**0.962**	0.758	0.704
Cropped Full Resolution nnUNet	0.971	0.852	0.812	0.947	0.752	0.707
BANET	0.970	0.840	0.798	0.949	0.740	0.694
Uncertainty Multi Checkpoint (Best 5)	0.971	**0.865**	**0.835**	0.945	**0.767**	**0.737**

2.3 Evaluation Metric

The Dice similarity coefficient (DSC) and Surface DCS (SD) are used as the main evaluation metrics. The mean values of these metrics are reported.

2.4 Results

Table 1 summarizes the segmentation performance of our models. In terms of Dice similarity coefficient (DSC), the cascaded nnUNet architecture achieved a mean kidney DSC of 0.979, which is higher than the single-stage low-resolution nnUNet model (0.971) and the cropped full-resolution nnUNet model (0.971). It also performed better than the BANET model (0.970).

For mass segmentation, the ensemble of our uncertain multi-checkpoint model (0.865) showed a significant improvement over the single-stage low-resolution nnUNet model (0.843), the cascaded nnUNet model (0.854), the cropped full-resolution nnUNet model (0.852), and the BANET model (0.840).

In the tumor segmentation task, the uncertain multi-checkpoint ensemble (0.835) again outperformed all other models, including the low-resolution nnUNet model (0.781), the cascaded nnUNet model (0.802), the cropped full-resolution nnUNet model (0.812), and the BANET model (0.798).

With respect to Surface Dice (SD), the cascaded nnUNet model (0.962) exhibited superior performance in the kidney segmentation task when compared to other models. However, in the mass and tumor segmentation tasks, the uncertain multi-checkpoint ensemble model demonstrated better performance with scores of 0.767 and 0.737 respectively, outperforming all other models.

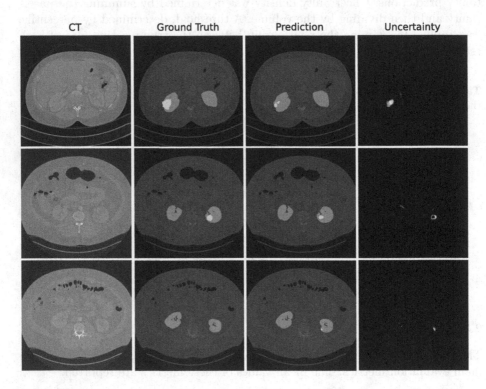

Fig. 2. Visualization of the CT scans, ground truth labels, predicted segmentation, and uncertainty maps.

Figure 2 shows the CT scan, corresponding prediction, and the associated uncertainty with the tumor prediction. The uncertainty predictions in rows 1 and 2 emphasize the regions where the predicted segmentation is absent, yet the ground truth exists. The third row demonstrates high uncertainty for a tumor prediction that turns out to be a false positive. Figure 3 depicts the true positive rate for both retained and excluded tumor and cysts after applying an uncertainty threshold determined by various percentiles.

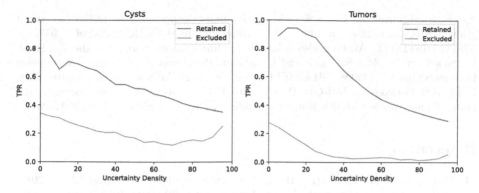

Fig. 3. The relationship between true positive rate and the volume of lesions after applying uncertainty threshold based on 95 percentile.

3   Discussion

In this study, we leveraged the low-resolution nnUNet to reduce the dimensionality of CT images. This strategic approach not only focused the network's attention on the crucial regions of the input image but also expedited the training and inference time compared to the cascaded nnUNet. Moreover, the introduction of a cyclic learning rate and posterior sampling of the weight space was helpful in quantifying the uncertainty of the predicted segmentation. We ensembled five models from different training cycles for our final model.

The ensemble model with uncertainty estimation outperformed other models in terms of the tumor and mass dice and surface dice metrics. In contrast, the cascaded nnUNet configuration outperformed in terms of the kidney dice and surface dice measurements. Furthermore, the predicted uncertainty maps indicated a positive correlation with the false positive detection. In future work, a post-processing step could be applied to eliminate false positives based on the uncertainty density of the masses.

4   Conclusion

In this research, we proposed a modified cascaded nnUNet framework for segmenting kidneys, kidney tumors, and kidney cysts in CT scans, which employs a cropping strategy to reduce data dimensionality and uncertainty estimation. Our approach demonstrated superior performance, particularly in the tumor and mass segmentation tasks, outperforming other tested models, such as the standard low-resolution nnUNet, the cascaded nnUNet, and the BANet. Moreover, our ensemble model, utilizing models from different training cycles, showed promising potential for detecting false positives.

Acknowledgements. We acknowledge financial support from ERC advanced grant (ERC-ADG-2015 n° 694812 - Hypoximmuno), ERC-2020-PoC: 957565-AUTO.DISTINCT. Authors also acknowledge financial support from the European Union's Horizon 2020 Research and Innovation Programme under grant agreement: ImmunoSABR n° 733008, CHAIMELEON n° 952172, EuCanImage n° 952103, JTI-IMI2-2020-23-two-stage IMI-OPTIMA n° 101034347. This work was supported by the Dutch Cancer Society (KWF Kankerbestrijding), Project number: 14449/2021-PoC.

References

1. Ferlay, J., et al.: Estimating the global cancer incidence and mortality in 2018: Globocan sources and methods. Int. J. Cancer **144**(8), 1941–1953 (2019)
2. Heller, N., et al.: The state of the art in kidney and kidney tumor segmentation in contrast-enhanced CT imaging: results of the KiTS19 challenge. Med. Image Anal. **67**, 101821 (2021)
3. Heller, N., et al.: The KiTS21 challenge: automatic segmentation of kidneys, renal tumors, and renal cysts in corticomedullary-phase CT. arXiv preprint arXiv:2307.01984 (2023)
4. Hu, S., Liao, Z., Ye, Y., Xia, Y.: Boundary-aware network for kidney parsing. In: Xiao, Y., Yang, G., Song, S. (eds.) CuRIOUS KiPA MELA 2022. LNCS, vol. 13648, pp. 9–17. Springer, Cham (2022). https://doi.org/10.1007/978-3-031-27324-7_2
5. Isensee, F., Jaeger, P.F., Kohl, S.A., Petersen, J., Maier-Hein, K.H.: nnU-net: a self-configuring method for deep learning-based biomedical image segmentation. Nat. Methods **18**(2), 203–211 (2021)
6. Jungo, A., Reyes, M.: Assessing reliability and challenges of uncertainty estimations for medical image segmentation. In: Shen, D., et al. (eds.) MICCAI 2019, Part II. LNCS, vol. 11765, pp. 48–56. Springer, Cham (2019). https://doi.org/10.1007/978-3-030-32245-8_6
7. Lambin, P., et al.: Radiomics: the bridge between medical imaging and personalized medicine. Nat. Rev. Clin. Oncol. **14**(12), 749–762 (2017)
8. de Leon, A.D., Pedrosa, I.: Imaging and screening of kidney cancer. Radiol. Clin. **55**(6), 1235–1250 (2017)
9. Minaee, S., Boykov, Y., Porikli, F., Plaza, A., Kehtarnavaz, N., Terzopoulos, D.: Image segmentation using deep learning: a survey. IEEE Trans. Pattern Anal. Mach. Intell. **44**(7), 3523–3542 (2021)
10. Salahuddin, Z., et al.: From head and neck tumour and lymph node segmentation to survival prediction on PET/CT: an end-to-end framework featuring uncertainty, fairness, and multi-region multi-modal radiomics. Cancers **15**(7), 1932 (2023)

An Ensemble of 2.5D ResUnet Based Models for Segmentation of Kidney and Masses

Cancan Chen[1] and Rongguo Zhang[1,2]($\boxtimes$)

[1] Infervision Advanced Research Institute, Beijing, China
[2] Academy for Multidisciplinary Studies, Capital Normal University, Beijing, China
zrongguo@cnu.edu.cn

Abstract. The automatic segmentation of kidney, kidney tumor and kidney cyst on Computed Tomography (CT) scans is a challenging task due to the indistinct lesion boundaries and fuzzy texture. Considering the large range and unbalanced distribution of CT scans' thickness, 2.5D ResUnet is adopted to build an efficient coarse-to-fine semantic segmentation framework in this work. A set of 489 CT scans are used for training and validation, and an independent never-before-used CT scans for testing. Finally, we demonstrate the effectiveness of our proposed method. The dice values on test set are 0.954, 0.792, 0.691, the surface dice values are 0.897, 0.591, 0.541 for kidney, tumor and cyst, respectively. The average inference time of each CT scan is 20.65 s and the max GPU memory is 3525 MB. The results suggest that a better trade-off between model performance and efficiency.

Keywords: Coarse-to-fine · Semantic-segmentation · ResUnet · KiTS23

1 Introduction

In recent years, over 430,000 people are diagnosed with kidney cancer and roughly 180,000 deaths are caused by kidney cancer annually [10]. Kidney tumors are found in an even larger number each year, and in most circumstances, it's not currently possible to radiographically determine whether a given tumor is malignant or benign [1]. Computer tomography (CT) scans is an import clinical tool to diagnose and detect kidney tumors. Surgery is the most common treatment option. Radiologists and surgeons are also dedicated to study kidney tumors on CT scans to design optimal treatment schedule by annotating the kidney and its masses manually. However, the manual annotation is a repetitive heavy laborious work and always subjective and varied from the different radiologists. Considering this, automatic segmentation of kidney and kidney tumors is a promising tool for alleviating these clinical problems.

Based on the 2019 and 2021 Kidney Tumor Segmentation Challenge [3,4], KiTS23 features an expanded training set (489 cases) with a fresh never-before-used test set (110 cases), and aims to serve a stronger benchmark and develop

© The Author(s), under exclusive license to Springer Nature Switzerland AG 2024
N. Heller et al. (Eds.): KiTS 2023, LNCS 14540, pp. 47–53, 2024.
https://doi.org/10.1007/978-3-031-54806-2_7

the best automatic semantic segmentation system for kidney tumors. Besides, hardware (GPU, CPU, etc.)about average inference time of each case are also real factors in clinical application scenes, so it is important to balance the performance and efficiency of the automatic semantic segmentation system.

In this paper, based on the original ResUnet [2], we propose an efficient coarse-to-fine semantic segmentation framework to automatically segment kidneys and tumors. In the coarse segmentation stage, the whole CT images are re-sampled to $128 \times 128 \times 128$ as the input. In the fine segmentation stage, we firstly obtain regions of interest (ROIs) for the kidney on the whole CT images based on the coarse segmentation mask, and according to this, randomly crop cubes along z-axis, which are re-sampled to $48 \times 224 \times 384$ as the input. Besides, a cascaded model, consisting of the kidney segmentation model and the kidney-tumor-cyst segmentation model, is applied on the second stage.

The main contributions of this work are summarized as follows:

- We propose a coarse-to-fine semantic segmentation framework, which can effectively segment kidney, kidney tumor and kidney cyst from the abdominal CT images.
- We firstly conduct a statistical analysis on the spacing resolution of all CT images, especially the thickness distribution at the z-axis, which sparks the major design ideas about the random cropping method, patch size and 2.5D ResUnet structure on the fine segmentation stage.
- We evaluate our proposed framework by 5-fold cross validation on Kits23 data set.

2 Methods

Semantic segmentation of organs and lesions is a common task for medical image analysis. There are already numerous accurate and efficient algorithms for medical image segmentation, such as U-Net [9], ResUNet [2], nnU-Net [5], et al. Based on the natural properties of Kits23 CT images and the strong baseline [3], we develop a whole-volume-based coarse-to-fine framework as follows, which consists of coarse segmentation, fine kidney segmentation (two-classification task of kidney and others) and fine tumor-mass segmentation (three-classification task of tumor, cyst and other kidney regions).

2.1 Preprocessing

Our proposed method includes the following preprocessing steps:

- Cropping strategy:

  In the coarse segmentation stage, the input is the whole volumes. In the fine segmentation stage, the kidney ROIs are firstly cropped from the whole volumes based on the coarse segmentation mask, and after that, we randomly crop 3D cubes from the kidney ROIs only along z-axis to ensure 2D kidney scans' structural integrity.

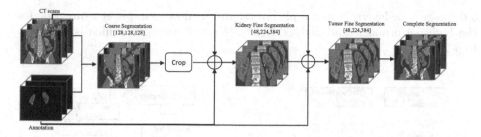

Fig. 1. An overview of our coarse-to-fine segmentation framework.

- Re-sampling method for anisotropic data:
  The original images are re-sampled to $128 \times 128 \times 128$ for coarse segmentation. In the fine segmentation stage, if the shape of the cropped kidney ROI is $d \times w \times h$, it will be resampled to $d \times 224 \times 384$ (in this work, $d = 48$), i.e., no-re-sampling at z-axis direction, and re-sampling at x/y-axis direction due to the shape distribution of all kidneys.
- Intensity normalization method:
  Images are clipped to range $[-200, 400]$ and normalized to range $[-1, 1]$.
- Others:
  To improve the training and testing efficiency, mixed precision is adopted in the whole process of our framework working.

2.2 Proposed Method

Our proposed framework is shown in Fig. 1. The details of two stages are addressed as follows.

Coarse Segmentation. We firstly use a original ResUnet [2] to obtain the coarse segmentation mask of all kidneys, and the input size is $128 \times 128 \times 128$. The kidney tumor and masses are always located in the kidney region. Based on this, the kidney ROI of each CT image is cropped as the input of the next segmentation stage. This step reduces the computational cost of irrelevant information on this task and preserves all segmentation target.

Fine Segmentation. The fine segmentation consists of kidney fine segmentation and lesion fine segmentation. Notably, the thickness range of all CT scans is between 0.5 mm and 5 mm. To resolve the data heterogeneity, cropping or re-sampling should be used. Considering the framework efficiency, the kidney ROIs are re-sampled to the fixed size at x and y direction, and then, we crop the cubes from kidney ROIs only along z axis. That's to say, if the shape of the kidney ROI is $d \times w \times h$, it will be re-sampled to $d \times 224 \times 384$, and the re-cropped cube size for fine segmentation is $48 \times 223 \times 384$ in this work. Finally, we adopt 2.5D ResUnet as the segmentation backbone. Network architecture has

3 down-sample layers, 3 up-sample layers, and no down-sample at z direction for the high-performance and high-efficiency of our framework, which is shown in Fig. 2.

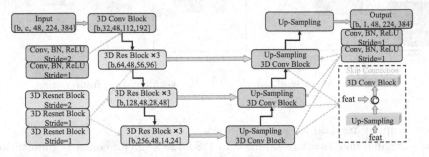

Fig. 2. Our proposed network architecture.

Loss Function. We use the summation of the weighted Dice loss and Cross-Entropy loss as the final compound loss function which has been proved to be robust in various medical image segmentation tasks [7].

Other Tricks. The mixup [12] and hard examples mining are adopted in the model training process, both of which significantly improve the ResUnet's fitting capability.

2.3 Post-processing

In the inference process, the connected component analysis [11] is applied to avoid the influence of noise. Based on the natural attributes of the kidney and lesions, we choose connected component regions larger than 10000 pixels as the final segmentation results. Notably, we abandon the multi-models ensemble method for efficient inference. Our method consists of the coarse segmentation model, the kidney fine segmentation model (background, kidney) and the lesion fine segmentation model (kidney, cyst, tumor). The final result is the average of the two predictions for the original image and the mirror image along the z-axis.

3   Results

3.1   Dataset and Evaluation Measures

The KiTS23 organizer has publicly released an expanded training set, totally 489 cases, based on KiTs19 and KiTs21. The volumetric Dice coefficient and the Surface Dice are used to evaluate algorithms, and the following Hierarchical Evaluation Classes (HECs) will be used: Kidney + Tumor + Cyst, Tumor + Cyst and Tumor only.

3.2   Implementation Details

Environment Settings. The development environments and requirements are presented in Table 1.

Table 1. Development environments and requirements.

Windows/Ubuntu version	Ubuntu 18.04.06 LTS
CPU	Intel(R) Core(TM) i9-10900X CPU @ 3.70 GHz
RAM	96 GB
GPU (number and type)	Four NVIDIA RTX A4000 16G
CUDA version	11.5
Programming language	Python 3.7
Deep learning framework	Pytorch (Torch 1.7.1+cu110, torchvision 0.8.2)
Specific dependencies	
(Optional) Link to code	

Training Protocols. In our training process, we performed the following data augmentation with project MONAI [8]: 1) randomly crop the volumes with range $[0.6, 1.3]$; 2) add brightness, contrast and gamma augmentation on the volumes and lesions with range $[0.6, 1.5]$, respectively. 3) random elastic transform with prob=0.5 and with sigma from range 3 to 5 and magnitude from range 100 to 200; 4) clip volumes to range $[-1, 1]$. Details of our training protocols are shown in Table 2 and Table 3.

Table 2. Training protocols for coarse segmentation.

Network initialization	"he" normal initialization
Batch size	4
Patch size	$128 \times 128 \times 128$
Total epochs	300
Optimizer	ADAMW [6] ($weightdecay = 1e - 4$)
Initial learning rate (lr)	1e−4
Lr decay schedule	CosineAnnealing

Table 3. Training protocols for fine segmentation.

Network initialization	"he" normal initialization
Batch size	4
Patch size	48 × 224 × 384
Total epochs	600
Optimizer	ADAMW [6] ($weightdecay = 1e - 4$)
Initial learning rate (lr)	1e−4
Lr decay schedule	CosineAnnealing

3.3 Results on Cross Validation and Test Data

Originally, our proposed framework would be evaluated by 5-fold cross validation. However, we only train and evaluate model on fold-0 data set due to the time and computation resource constraints, and all scores are listed in Table 4. The average inference time on fold-0 validation (98 cases) and test set (110 cases) is 19.22 s and 20.65 s, respectively. The max GPU memory at inference step is 3525 MB.

Table 4. Results of our proposed method on fold-0 and test set.

Targets	Dice (Fold-0)	Surface Dice (Fold-0)	Dice (Test)	Surface Dice (Test)
Kidney	0.9661	0.9408	0.954	0.897
Cyst	0.8580	0.7365	0.792	0.591
Tumor	0.8591	0.7329	0.691	0.541

4   Conclusion

Based on 2.5D ResUnet, we propose a efficient coarse-to-fine framework for the automatic segmentation of kidney and masses. The experimental results indicate that our framework is effective, but the segmentation robustness of kidney tumors and cysts need further improvement. One possible reason is that the capability level of single model has a lower upper-limit for the hard segmentation task. Thus, the ensemble of multi-models is an alternative solution after balancing performance and efficiency.

References

1. De, L., Alberto, D., Ivan, P.: Imaging and screening of kidney cancer. Radiol. Clin. **55**, 1235–1250 (2017)
2. Diakogiannis, F.I., Waldner, F., Caccetta, P., Wu, C.: ResUnet-a: a deep learning framework for semantic segmentation of remotely sensed data. ISPRS J. Photogramm. Remote. Sens. **162**, 94–114 (2020)
3. Heller, N., et al.: The state of the art in kidney and kidney tumor segmentation in contrast-enhanced CT imaging: results of the KiTS19 challenge. Med. Image Anal. **67**, 101821 (2021)
4. Heller, N., et al.: An international challenge to use artificial intelligence to define the state-of-the-art in kidney and kidney tumor segmentation in CT imaging. Proc. Am. Soc. Clin. Oncol. **38**(6), 626 (2020)
5. Isensee, F., Jaeger, P.F., Kohl, S.A., Petersen, J., Maier-Hein, K.H.: nnU-net: a self-configuring method for deep learning-based biomedical image segmentation. Nat. Methods **18**(2), 203–211 (2021)
6. Loshchilov, I., Hutter, F.: Decoupled weight decay regularization. arXiv preprint arXiv:1711.05101 (2017)
7. Ma, J., et al.: Loss odyssey in medical image segmentation. Med. Image Anal. **71**, 102035 (2021)
8. Nic, M., Wenqi, L., Richard, B., Yiheng, W., Behrooz, H.: MONAI. https://github.com/Project-MONAI/MONAI. [Version 0.8.1]
9. Ronneberger, O., Fischer, P., Brox, T.: U-net: convolutional networks for biomedical image segmentation. In: Navab, N., Hornegger, J., Wells, W.M., Frangi, A.F. (eds.) MICCAI 2015. LNCS, vol. 9351, pp. 234–241. Springer, Cham (2015). https://doi.org/10.1007/978-3-319-24574-4_28
10. Sung, H., et al.: Global cancer statistics 2020: GLOBOCAN estimates of incidence and mortality worldwide for 36 cancers in 185 countries. CA Cancer J. Clin. **71**, 209–249 (2021)
11. William, S., Nico, K.: Odidev: CC3D. https://github.com/seung-lab/connected-components-3d. [Version 3.10.1]
12. Zhang, H., Cisse, M., Dauphin, N, Y., Lopez-Paz, D.: Mixup: beyond empirical risk minimization. arXiv preprint arXiv:1710.09412 (2017)

Using Uncertainty Information
for Kidney Tumor Segmentation

Joffrey Michaud$^{(\boxtimes)}$, Tewodros Weldebirhan Arega, and Stephanie Bricq

ImViA Laboratory, EA 7535, Universit de Bourgogne, Dijon, France
joffrey.michaud@u-bourgogne.fr

Abstract. Kidney cancer occurrence increases since 1990's and its main treatment is surgery. According to this, performing automatic segmentation is an important tool to develop. In this paper, we used a two stages pipeline to get the segmentation of kidney, tumor and cyst. The first stage is used to segment the kidney region to allow us to crop the data. The second stage leverages uncertainty using Monte-Carlo dropout during training by introducing an uncertainty estimate term in the loss function.

Keywords: Kidney semantic segmentation · 3D U-Net · Uncertainty

1 Introduction

Kidney cancer is the 14th most common cancer worldwide, 9th in men and 14th in women. With more than 430,000 new cases in 2020 its incidence is increasing since 1990's. Its main treatment is surgery. Therefore, segmentation is a very important step that can be laborious and time-consuming when performed manually. That is why automatic segmentation of kidney and kidney masses can be powerful and help practitioners in patient management. Deep learning is very helpful in this domain as we can see with the widely used nnUNet framework [6].

2 Methods

The open source framework nnUNet [6] is a very powerful tool to perform medical image segmentation massively used for this kind of tasks. Its efficiency has been demonstrated on several datasets and particularly it has shown very good results on kidney tumor segmentation [6]. We decided to use this framework to develop our segmentation pipeline in two stages inspired by the winner of the previous challenge KiTS21 [9].

The first stage uses a 3D U-net to segment kidney region in order to crop the data in smaller images. Then, the cropped images are fed into a second 3D U-net to get the final segmentation.

© The Author(s), under exclusive license to Springer Nature Switzerland AG 2024
N. Heller et al. (Eds.): KiTS 2023, LNCS 14540, pp. 54–59, 2024.
https://doi.org/10.1007/978-3-031-54806-2_8

In addition to this we used a loss function with an additional term to leverage uncertainty during training, using Monte-Carlo dropout (MC-dropout).

2.1 Training and Validation Data

Our submission made use of the official KiTS23 training set alone, composed of 489 cases. The semantic labels are kidney, tumor, and cyst. We performed a 5-fold cross validation to evaluate our method during training. All following results are means over the five folds used during the training.

2.2 Preprocessing

We followed the baseline process of nnUNet to preprocess the dataset. A variety of data augmentation is applied on the fly during training like rotation, scaling, low resolution simulation, etc.

2.3 Proposed Method

Our method is build in two stages. The first segments the kidney region to allow data to be cropped around kidney region. Then we trained a second network with the cropped dataset to get the final segmentation.

Cropping Step. The first stage uses a conventional 3D U-Net network to segment only the kidney area. Then we cropped the dataset with small boundaries around the kidney segmentation. This model is trained during 1000 epochs on Nvidia GPUs V100, using a batch-size of 2. 3D convolutions of the network was performed with $3 \times 3 \times 3$ kernel.

Final Segmentation. To get the final segmentation of all needed structures, we trained a 3D U-Net fed with the cropped dataset. This network is trained following the MC-dropout method. The 3D U-Net network is slightly modified by adding dropout layers after the four middle stages, according to literature [4,5,7,8]. These dropout layers allow us to sample the network N times during training, and to compute uncertainty [1–3] estimate to add to the loss function.

Uncertainty Loss. According to MC-dropout method, the network is trained with dropout layers. During training, the network is sampled N times and the mean of the N segmentations is used as final segmentation.

The loss function is a combination of segmentation loss and uncertainty estimate (Eq. 1), where α is an hyper-parameter that controls the contribution of uncertainty to the final loss. The segmentation loss is a weighted sum of Dice and Focal loss (Eq. 2).

$$L_{Total} = L_{Seg} + \alpha \times L_{Uncertainty} \tag{1}$$

$$L_{Seg} = \lambda_{Dice} L_{Dice} + \lambda_{Focal} L_{Focal} \tag{2}$$

The uncertainty estimate used in the loss function is computed from the N MC-dropout samples. Each pixel i has predictions $(y_1, \ldots, y_N)$, from which we can compute pixel-wise mean μ_i (3) and variance σ_i^2 (4). To compute image-level uncertainty, the per-pixel uncertainty is averaged over all pixels in the image (Eq. 5).

$$\mu_i = \sum_{n=1}^{N} (y_{i,n}) \tag{3}$$

$$\sigma_i^2 = \frac{1}{N} \sum_{n=1}^{N} (y_{i,n} - \mu_i)^2 \tag{4}$$

$$L_{Uncertainty} = \frac{1}{I} \sum_{i=1}^{I} (\sigma_i^2) \tag{5}$$

During inference, dropout layers stay activated and the segmentation output of the network is the average of the N samples (Fig. 1).

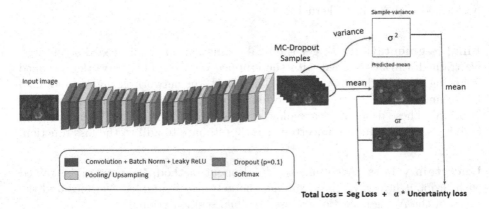

Fig. 1. U_Net with additional dropout layers for MC-dropout

3   Results

We evaluate the results of our different losses with the Dice coefficient metric, averaged over the 5 folds of the 5-fold cross validation. This score is computed using Hierarchical Evaluation Classes (HEC) for the three classes (Kidney, Tumor, Cyst) defined as follow: the first class, Kidney and Masses, includes Kidney, Tumor and Cyst. The second class is Kidney Mass and includes Tumor and Cyst. The last class is only Tumor.

In this paper, we compare nnUNet Baseline (Dice and Cross-Entropy), Focal loss, weighted sum of Focal and Dice, and same losses with an additional term representing uncertainty contribution. Results are shown in Table 1.

Our method achieved the 10th position for the KiTS23 challenge on the official test set. The detailed scores can be found in Table 2.

Table 1. Segmentation performance in term of Dice score. The bold results are better.

Method	Kidney	Masses	Tumor
Baseline	0.9651	0.8361	0.7959
Focal	0.9652	0.818	0.7655
Dice+Focal	0.9604	0.8363	0.7891
Proposed (Focal+Uncertainty)	0.9658	0.8156	0.7594
Proposed (Dice+Focal+Uncertainty)	0.9601	0.8356	0.7929
Proposed (Baseline+Uncertainty)	**0.9659**	**0.8394**	**0.8018**

Table 2. Official KiTS23 results on test set

	Average	Tumor	Masses	Kidney and Masses
Dice	0.790	0.670	0.750	0.949
Surface Dice	0.678	0.531	0.603	0.899

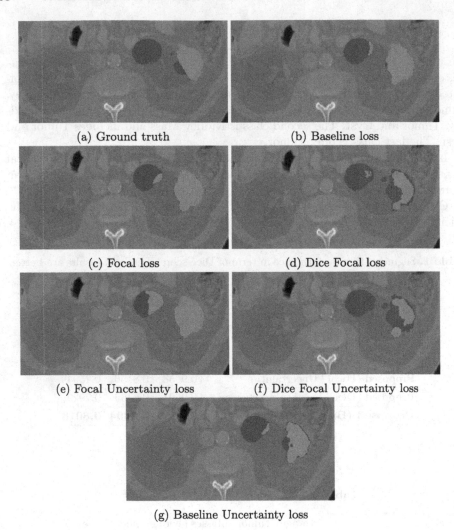

(a) Ground truth   (b) Baseline loss

(c) Focal loss   (d) Dice Focal loss

(e) Focal Uncertainty loss   (f) Dice Focal Uncertainty loss

(g) Baseline Uncertainty loss

Fig. 2. Segmentation examples on the different losses

4   Discussion and Conclusion

The Table 1 shows results for different tested methods. The best scores are achieved using a combination of the nnUNet baseline loss (Dice loss and Cross-Entropy) for the segmentation loss, and adding a term representing the uncertainty level for the segmentation using MC-dropout. The weight α has been empirically choose with the value 1.0 (Fig. 2).

This method improves by a little the nnUNet baseline and shows that using uncertainty estimate can improve the results. This information could be useful in clinical practice. For future works, it could be interesting to integrate to integrate the uncertainty information in the post-processing step in order to improve the final segmentation performance.

References

1. Arega, T.W., Bricq, S., Legrand, F., Jacquier, A., Lalande, A., Meriaudeau, F.: Automatic uncertainty-based quality controlled T1 mapping and ECV analysis from native and post-contrast cardiac T1 mapping images using Bayesian vision transformer. Med. Image Anal. **86**, 102773 (2023). https://doi.org/10.1016/j.media.2023.102773
2. Arega, T.W., Bricq, S., Meriaudeau, F.: Leveraging uncertainty estimates to improve segmentation performance in cardiac MR. In: Sudre, C.H., et al. (eds.) UNSURE/PIPPI -2021. LNCS, vol. 12959, pp. 24–33. Springer, Cham (2021). https://doi.org/10.1007/978-3-030-87735-4_3
3. Arega, T.W., Bricq, S., Meriaudeau, F.: Using polynomial loss and uncertainty information for robust left atrial and scar quantification and segmentation. In: Zhuang, X., Li, L., Wang, S., Wu, F. (eds.) LAScarQS 2022. LNCS, vol. 13586, pp. 133–144. Springer, Cham (2022). https://doi.org/10.1007/978-3-031-31778-1_13
4. Blundell, C., Cornebise, J., Kavukcuoglu, K., Wierstra, D.: Weight uncertainty in neural network. In: International Conference on Machine Learning, pp. 1613–1622. PMLR (2015)
5. Fortunato, M., Blundell, C., Vinyals, O.: Bayesian recurrent neural networks (2017). http://arxiv.org/abs/1704.02798
6. Isensee, F., Jaeger, P.F., Kohl, S.A.A., Petersen, J., Maier-Hein, K.H.: nnU-net: a self-configuring method for deep learning-based biomedical image segmentation. Nat. Methods **18**(2), 203–211 (2021). https://doi.org/10.1038/s41592-020-01008-z
7. Kendall, A., Badrinarayanan, V., Cipolla, R.: Bayesian SegNet: model uncertainty in deep convolutional encoder-decoder architectures for scene understanding (2015). http://arxiv.org/abs/1511.02680
8. Ng, M., et al.: Estimating uncertainty in neural networks for cardiac MRI segmentation: a benchmark study. IEEE Trans. Biomed. Eng. **70**(6), 1955–1966 (2022). https://doi.org/10.1109/TBME.2022.3232730
9. Zhao, Z., Chen, H., Wang, L.: A coarse-to-fine framework for the 2021 kidney and kidney tumor segmentation challenge. In: Heller, N., Isensee, F., Trofimova, D., Tejpaul, R., Papanikolopoulos, N., Weight, C. (eds.) KiTS 2021. LNCS, vol. 13168, pp. 53–58. Springer, Cham (2022). https://doi.org/10.1007/978-3-030-98385-7_8

Two-Stage Segmentation and Ensemble Modeling: Kidney Tumor Analysis in CT Images

Soohyun Lee[1(✉)], Hyeyeon Won[1], and Yeeun Lee[2]

[1] Department of Electrical and Electronic Engineering, Yonsei University,
Seoul, Republic of Korea
suhyun2374@yonsei.ac.kr
[2] Department of Artificial Intelligence, Yonsei University, Seoul, Republic of Korea

Abstract. In the realm of kidney cancer, accurate segmentation is pivotal for effective diagnosis and treatment. Participating in the 2023 KiTS Challenge as a platform, our research introduces a two-stage strategy combining the strengths of nnU-Net and nnFormer for enhanced tumor segmentation. Our approach focuses on the kidney region, facilitating the learning of tumor-influenced areas, and employs an ensemble of two nnU-Net models for precise segmentation. Evaluated on the KiTS23 dataset, which emphasizes the segmentation of the kidney, tumor, and cyst, our method demonstrated its potential in addressing complex medical image segmentation challenges.

Keywords: Automatic kidney segmentation · Two-stage framework · Model ensembling

1 Introduction

Recently, advancements in deep neural networks have significantly accelerated research [3,7,8] in the segmentation of CT and MRI medical images. Kidney-related diseases, especially tumors and cysts, have been a significant concern in the medical community. Accurate segmentation of the kidney, its tumors, and cysts is crucial for effective diagnosis and treatment. With the advent of deep learning techniques, there has been an increase in the development of automated tools that can assist radiologists in segmenting medical images with high precision. Among these techniques, the U-Net [6] based segmentation methods, particularly nnU-Net [2], have shown promising results in various medical image segmentation tasks. Alongside these, transformer-based [1] methods, especially nnFormer [10], are also being extensively researched and have begun to demonstrate their potential in medical image segmentation tasks. However, with the increasing complexity of medical images and the need for accurate segmentation, there's a growing demand for more advanced and hybrid models [4].

© The Author(s), under exclusive license to Springer Nature Switzerland AG 2024
N. Heller et al. (Eds.): KiTS 2023, LNCS 14540, pp. 60–66, 2024.
https://doi.org/10.1007/978-3-031-54806-2_9

In this context, the 2023 KiTS Challenge provides an excellent platform for researchers to showcase and test their innovative methods in tumor segmentation. Our paper presents an approach that combines the strengths of both nnU-Net and nnFormer. We utilize nnFormer, capable of capturing foreground objects without omitted regions, and nnU-Net, which excels in finer segmentation, at different stages. By combining the two models, we anticipate achieving more precise segmentation of regions. Harnessing the advantages of these models in a two-stage strategy, we aim to achieve superior segmentation performance in the challenging task of tumor segmentation. Our research provides various contributions by improving existing models in the following ways:

1. We propose a two-stage method aimed at focusing on a model that focuses more on learning the kidney and masses(tumor, and cyst). This approach enabled the model to effectively learn the representation of the tumor within the kidney.
2. In the stage 2, we used ensembling approach with two models employing nnU-Net, including whole CT image and tumor only segmentation model. Significantly, the combination of this tumor-focused model led to an enhancement in tumor dice similarity coefficient (DSC) for tumor segmentation.

The KiTS23 challenge is a segmentation competition that involves segmenting three classes: kidney, tumor, and cyst. In this challenge, this includes the DSC and surface dice (SD) for the three hierarchical evaluation classes (HEC)'s kidney and masses, masses and tumor. Upon evaluating these metrics, our team achieving the 11th place overall.

2   Methods

This study was inspired by nnU-net and nnFormer, The nnU-Net is a U-Net based segmentation method that provides automatically optimized configurations for various datasets and tasks. On the other hand, nnFormer is a transformer-based segmentation approach that leverages an empirical integration of self-attention and convolution within a cross-architecture. Using nnFormer and nnU-net, we cropped the RoI regions and devised a method to selectively combine two models yielding the outstanding performance by exploring their performance on the tumor segmentation. The proposed method is visualized in Fig. 1.

2.1   Training and Validation Data

Our submission made use of the official KiTS23 training set alone.

2.2   Preprocessing

To preprocess the training data, we carry out several preprocessing steps. We firstly crop the non-zero regions to extract a clear foreground. Then we normalized CT images which is important to medical image processing [5]. To uniformly convert all CT data to a common spatial resolution, we recognize the shape of the original data and the original spatial resolution, and calculate a new data shape based on the target spatial resolution of target voxel spacing [3.0, 0.78125, 0.78125]. The new shape of each axis is obtained by multiplying the ratio of the original spatial resolution to the target spatial resolution by the number of pixels in the original axis, and resampling is performed accordingly.

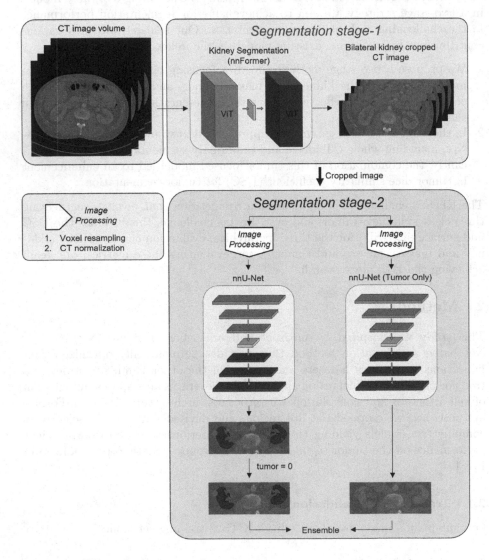

Fig. 1. An overview of our two-stage segmentation framework

2.3   Proposed Method

The proposed method consists of two-stage, employing nnFormer in the first stage and nnU-Net in the second stage.

Stage 1. Coarse segmentation for RoI

In the Stage 1 of the method, we used nnFormer trained on the entire dataset to perform a coarse segmentation for using crop the foreground region [9]. Then, we utilized a mass center crop mechanism in our hard coding to crop from the original data in such a way that bilateral kidneys are included. For adequate learning, we set a buffer of 10 slices on both sides of the axis. This stage is essential to ensure that the learning focuses on the foreground regions.

Post-processing. The post-processing approach is as follows, counting the number of voxels in connected components and retaining only the two components with the highest voxel counts, while deleting the rest. Post-processing is used exclusively to first stage, aiming for a more precise and detailed cropping of the kidney region. We refrain from using post-processing in second stage due to the potential risk of erroneously removing the regions of interest.

Stage 2. Improved Segmentation via Dual nnU-Net Ensemble

We employ two nnU-Net models trained on cropped images. One model segments kidney, tumor, cyst, and background regions, while the other segments tumor and background. This way is expected to benefit from the focused training approach on the tumor area, a key aspect of tumor diagnosis.

The integration of a dedicated tumor-focused nnU-Net model into our ensemble approach leads to enhanced tumor segmentation. The specialized model's contribution enhances the Tumor DSC, underscoring the benefits of integrated models in ensemble setups for accurate medical image segmentation.

Ensembling Strategy. An ensembling strategy takes advantage of the individual strengths of the models while compensating for their individual weaknesses, thereby creating an ensemble model with superior predictive power. To better capture the most crucial tumor region, we used the tumor class from a model trained solely on the tumor, ignoring other foregrounds, instead of the tumor class from a model trained on 4-classes. When ensembling the models, we employ a weighted sum approach, using a ratio of 7:3 between the bilateral kidney cropped model and the tumor-only model to generate the final segmentation model.

Validation Strategy. The validation strategy is based on the 5-fold cross-validation method. The approach mitigates the risk of overfitting and to ensure that our models are trained and validated across the entire dataset. The dataset is split into 8:2 ratio for train and validation. After assessing the performance on the validation dataset, the fold that achieved the highest DSC was identified and selected as the optimal model.

Implementation Details. Our process is divided into two steps: using the nnForemr, nnU-net in stage 1, 2. We trained the nnFormer model with the convolution kernel set to $3 \times 3 \times 3$ and the pooling kernel set to $2 \times 2 \times 2$. In the nnU-Net, we applied a convolutional kernel of size $3 \times 3 \times 3$ and a pooling kernel of $1 \times 1 \times 1$ in the initial layer. For the subsequent layers, a pooling kernel of size $2 \times 2 \times 2$ was employed.

All models were trained using randomly sampled patches of size $128 \times 128 \times 128$ from the resampled volumes as input. Each model was trained for 1000 epochs using Stochastic Gradient Descent (SGD) as the optimizer, Instance Normalization to normalize the inputs across the layers, LeakyReLU as the activation function. with a batch size of 2 and 250 iterations per epoch. The models were trained to minimize a combined loss, which is the sum of cross-entropy and dice loss.

3 Results

We evaluated our proposed method using the KiTS23 dataset. In Table 1, we summarized the results for the kidney and masses, masses, and tumor based on the KiTS23 Test set for the following two metrics. Our Tumor Dice score stands at 0.697, which is on par with the team ranked 4th.

Table 1. Quantitative evaluation on the test dataset

	Kidney + Masses	Masses	Tumor
Dice	0.930	0.752	0.697
Surface Dice	0.874	0.597	0.548

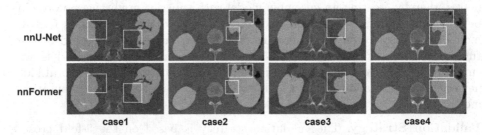

Fig. 2. Comparison of test set predictions between nnFormer and nnU-Net. It shows that nnFormer segments the wider the kidney and masses regions than nnU-Net.

4   Discussion and Conclusion

In this paper, we propose a framework which consists of two stage framework. In the first stage, we take advantage of nnFormer to extract the RoI using the segmentation results. In the second stage, we train nnU-Net with the bilateral kidney cropped data. At this juncture, we perform an ensemble of two models: one nnU-Net model trained on the 4-class dataset and another specifically trained considering only the tumor as the foreground.

The reason for using different models at each stage is to leverage the strengths of two distinct models. In our experiments, we find that nnFormer captures foreground objects in a rough manner compared to nnU-Net. We judge this to be unsuitable for precisely segmenting the kidney, tumor, and cyst in stage 2. Figure 2 aptly illustrates this observation. In this manner, we devised an optimal network suited for hierarchical evaluation. We hope that our research will contribute to future advancements in kidney cancer segmentation.

Acknowledgements. This research was supported by Basic Science Research Program through the National Research Foundation of Korea funded by the Ministry of Science and ICT (2022R1A2C2008983), Artificial Intelligence Graduate School Program at Yonsei University [No. 2020-0-01361], the KIST Institutional Program (Project No.2E32271-23-078), and partially supported by the Yonsei Signature Research Cluster Program of 2023 (2023-22-0008).

References

1. Dosovitskiy, A., et al.: An image is worth 16x16 words: transformers for image recognition at scale. arXiv preprint arXiv:2010.11929 (2020)
2. Isensee, F., Jaeger, P.F., Kohl, S.A., Petersen, J., Maier-Hein, K.H.: nnU-net: a self-configuring method for deep learning-based biomedical image segmentation. Nat. Methods **18**(2), 203–211 (2021)
3. Jun, Y., et al.: Intelligent noninvasive meningioma grading with a fully automatic segmentation using interpretable multiparametric deep learning. Eur. Radiol. 1–10 (2023)
4. Park, D., Jang, R., Chung, M.J., An, H.J., Bak, S., Choi, E., Hwang, D.: Development and validation of a hybrid deep learning-machine learning approach for severity assessment of COVID-19 and other pneumonias. Sci. Rep. **13**(1), 13420 (2023)
5. Park, D., et al.: Importance of CT image normalization in radiomics analysis: prediction of 3-year recurrence-free survival in non-small cell lung cancer. Eur. Radiol. **32**(12), 8716–8725 (2022)
6. Ronneberger, O., Fischer, P., Brox, T.: U-net: convolutional networks for biomedical image segmentation. In: Navab, N., Hornegger, J., Wells, W., Frangi, A. (eds.) MICCAI 2015 Part III 18. LNCS, vol. 9351, pp. 234–241. Springer, Cham (2015). https://doi.org/10.1007/978-3-319-24574-4_28
7. Shin, H., Kim, H., Kim, S., Jun, Y., Eo, T., Hwang, D.: SDC-UDA: volumetric unsupervised domain adaptation framework for slice-direction continuous cross-modality medical image segmentation. In: Proceedings of the IEEE/CVF Conference on Computer Vision and Pattern Recognition, pp. 7412–7421 (2023)

8. Shin, Y., et al.: Digestive organ recognition in video capsule endoscopy based on temporal segmentation network. In: Wang, L., Dou, Q., Fletcher, P.T., Speidel, S., Li, S. (eds.) MICCAI 2022. LNCS, vol. 13437, pp. 136–146. Springer, Cham (2022). https://doi.org/10.1007/978-3-031-16449-1_14
9. Zhao, Z., Chen, H., Wang, L.: A coarse-to-fine framework for the 2021 kidney and kidney tumor segmentation challenge. In: Heller, N., Isensee, F., Trofimova, D., Tejpaul, R., Papanikolopoulos, N., Weight, C. (eds.) KiTS2021. LNCS, vol. 13168, pp. 53–58. Springer, Cham (2021). https://doi.org/10.1007/978-3-030-98385-7_8
10. Zhou, H.Y., Guo, J., Zhang, Y., Yu, L., Wang, L., Yu, Y.: nnformer: Interleaved transformer for volumetric segmentation. arXiv preprint arXiv:2109.03201 (2021)

GSCA-Net: A Global Spatial Channel Attention Network for Kidney, Tumor and Cyst Segmentation

Xiqing Hu[1] and Yanjun Peng[1,2(✉)]

[1] College of Computer Science and Engineering, Shandong University of Science and Technology, No. 579, Qianwan'gang Road, Qingdao 266590, China
pengyanjuncn@163.com
[2] Shandong Province Key Laboratory of Wisdom Mining Information Technology, No. 579, Qianwan'gang Road, Qingdao 266590, China

Abstract. Automatic segmentation of the kidney, tumor, and cysts is crucial for the treatment of renal cancer. In this paper, we employed a 3D residual U-Net architecture as the pre-processing method to extract the region of interest (ROI) and segment the kidney. Then, we propose Global Spatial Channel Attention Network (GSCA-Net) with global spatial attention (GSA) and global channel attention (GCA) for the segmentation of tumors and cysts. Global spatial attention improves the global spatial representation ability, and global channel attention learns features between different channels. The GSCA module enhances the segmentation accuracy of tumors and cysts through the fusion of two parallel global attention modules. Furthermore, we employ a novel boundary loss function in GSCA-Net to improve the Surface Dice. On the official test set including cases 589–698, our approach achieves Dice coefficients of 0.933, 0.744, and 0.679 for the kidney, masses, and tumor, respectively.

Keywords: Global channel attention · Global spatial attention · Kidney tumor and cyst segmentation

1 Introduction

According to statistics, the annual number of deaths attributed to the kidney tumors exceeds 140,000 individuals [1]. Computed tomography (CT) images play a crucial role in detecting renal tumors and making accurate diagnoses. KiTS23 introduced nephrogenic contrast phase kidney CT data for image segmentation for the first time. In addition to the late arterial cases in KiTS2021, KiTS2023 also includes cases from the nephrogenic contrast phase of renal imaging. The key to distinguishing between renal tumors and renal cysts lies in the morphological differences observed in CT images. Cysts primarily result from fluid accumulation within the kidneys, forming fluid-filled sacs, while tumors are caused by the proliferation of cells [2].

© The Author(s), under exclusive license to Springer Nature Switzerland AG 2024
N. Heller et al. (Eds.): KiTS 2023, LNCS 14540, pp. 67–76, 2024.
https://doi.org/10.1007/978-3-031-54806-2_10

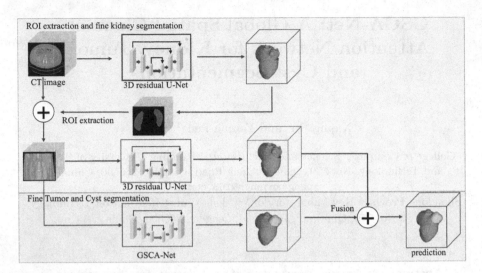

Fig. 1. Flow scheme of kidney and tumor segmentation method.

In the KiTS2019 [3], nnU-Net demonstrated strong adaptive capability and was validated to sufficiently segment kidney and tumor regions [4]. In the KiTS2021, Zhao et al. [5] employed four separate nnU-Net architectures and utilized a coarse-to-fine strategy for kidney, mass and tumor segmentation. Extracting ROIs helps avoid the network learning irrelevant region. By using multiple nnU-Net networks to segment the kidney, tumor and mass segmentation, the segmentation can be enhanced. The coarse-to-fine approach and the nnUNet structure have been proven to be highly effective for kidney and tumor segmentation [6,7]. Due to the similarity in appearance between some renal tumors and cysts, achieving precise segmentation of tumors and cysts are also challenging.

To this end, we use 3D residual U-Net to segment the ROIs for the kidneys. Within the ROIs, we further employ 3D residual U-Net for fine segmentation of kidney region. After two 3D residual U-Net, the segmentation of kidney has been very precise. Then using GSCA-Net can more accurately improve the segmentation accuracy of tumors and cysts. In our experiments, GSCA-Net needs to segment kidneys, tumors and cysts to get the best tumor and cyst results (kidney segmentation results in GSCA-Net are not good enough, so it needs to be fused with 3D residual U-Net's kidney results).

2   Methods

As shown in Fig. 1, in the pre-processing step, we employ 3D residual U-Net to generate the kidney region and extract the region of interest (ROIs) based on the kidney region. Then, we use 3D residual U-Net again within the ROIs to improve the kidney segmentation results. In the accurate segmentation results of

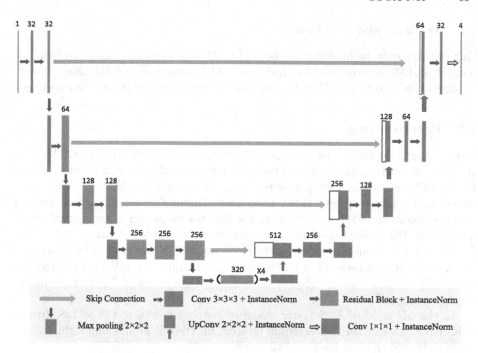

Fig. 2. 3D Residual U-Net for coarse and fine kidney segmentation.

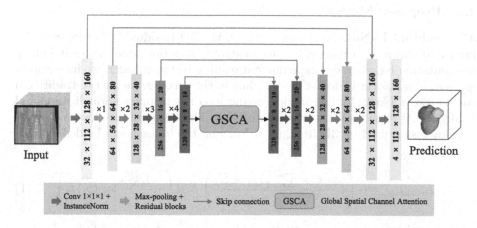

Fig. 3. 3D GSCA-Net for fine tumor and cyst segmentation.

the kidney, we adopted the Global Spatial Channel Attention Network (GSCA-Net) for fine kidney, tumor and cyst segmentation. Finally, we fuse the fine kidney segmentation results from 3D residual U-Net with the tumor and cyst segmentation from GSCA-Net to obtain the final segmentation result.

2.1   Training and Validation Data

Our dataset only include the official KiTS2023 dataset. 320 cases are training dataset and 80 cases are validation dataset. The training and validation set were randomly split from case ID 0 to 499. case ID from 500 to 588 are our test set.

2.2   Preprocessing

In each case, we follow the same preprocessing as nnUNet. In KiTS2023, The spacing varies in z-axis. We need a tradeoff between different segmentation steps. In the ROI extraction step, our objective is to segment the kidney region only, so we used a larger spacing [1.99, 1.99, 1.99]. In the fine segmentation step, to generate more detailed segmentation results, we employed a spacing of [0.78, 0.78, 0.78]. We applied B-spline interpolation to resample the data.

We applied intensity clipping to the 0.5% and 99.5% foreground intensity levels, with a clipping range of [-61, 309] Hounsfield Units (HU). We subtracted the mean value of 103.8 and divided by the standard deviation of 75.13 to normalize the input images.

In the 3D residual U-Net, our shape of patches is $128 \times 128 \times 128$. In the GSCA-Net, we extract the regions of interest for the left kidney and right kidney separately and the patch size is $112 \times 128 \times 160$.

2.3   Proposed Method

3D Residual U-Net. As shwon in Fig. 2, the 3D Residual U-Net [8], with its increased parameter capacity, has demonstrated superior performance in kidney segmentation based on our experimental results. In the encoder, we incorporate more residual blocks in deeper layers, while in the decoder, we use two traditional convolutional blocks for information extraction. The instance normalization and Leaky ReLU remain consistent with nnUNet. The fusion of dice loss and cross-entropy loss is used for 3D Residual U-Net and the formula can be defined as:

$$L_{residual} = \alpha \left(1 - \frac{C \times \sum_{i=1}^{C} (p_i \times t_i)}{\sum_{i=1}^{C} (p_i + t_i) + \theta} \right) + \beta \left(-\frac{1}{C} \sum_{i=1}^{C} \log (p_i) t_i \right) \quad (1)$$

where $p_i, t_i, i \in 0, 1$ is the prediction and ground truth (GT) of kidney region, θ is a small number to avoid division by zero and C equals 2. α and β are two learnable factors.

Proposed GSCA-Net. The architecture of GSCA-Net is shown in Fig. 3. We proposed a global attention block in GSCA-Net. Through our experiments, we have validated that the residual blocks enhances the precision of the network. Consequently, in our encoder architecture, similar to the 3D residual U-Net, we progressively increase the number of residual blocks into the encoder layers

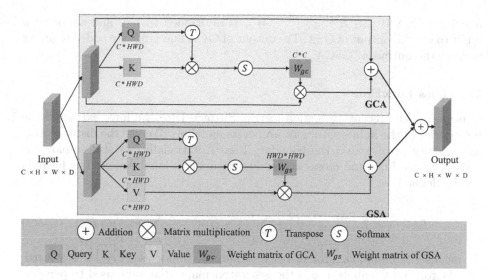

Fig. 4. Detail of Global Spatial Channel Attention block.

which is shown in Fig. 3. At the bottom layer of GSCA-Net, we integrate both global channel attention and global spatial attention in one module. This module significantly enhances the segmentation results of our network. To increase the parameter capacity of our network, each layer in our decoder consists of two residual blocks, which are consistent with the ones used in the encoder. Following this, we apply a $1 \times 1 \times 1$ 3D convolution to obtain the final prediction results. Next, we will focus on introducing our GSCA (Global Spatial Channel Attention) module.

Global Spatial Channel Attention. Attention mechanisms have proven to be a key approach for improving image segmentation performance, with channel attention and spatial attention being two popular directions [9–11]. Recently, with the adoption of transformer-based methods in the field of image segmentation, global attention has emerged as a new research direction [12,13]. As shown in Fig. 4, GSCA mainly composed of two parts: A Global Channel Attention (GCA) module for channel information extraction and a Global Spatial Attention (GSA) for spatial information extraction.

In GCA module, the input is reshaped to generate query and key $Q, K \in R^{C*HWD}$. Q is transposed and multiplied with K. After softmax operation, we get weight matrix of GCA $W_{gc} \in R^{C*C}$. W_{gc} is multiplied with input and added with input to get the output of GCA.

In GSA module, the input is reshaped to generate query, key and value $Q, K, V \in R^{C*HWD}$. Q is transposed. Different with GCA, K is multiplied by transposed Q to get the spatial matrix. After softmax operation, we get weight matrix of GSA $W_{gs} \in R^{HWD*HWD}$. W_{gs} is multiplied with V and added with input to get the output of GSA. The output of GCA and GSA are added together and get the output of GSCA.

2.4   Loss Function

Considering the recent evaluation metric, Surface Dice, we have extracted the boundaries of the kidneys, tumors, and cysts as labels. We use the distance map loss penalty to constrain the boundary loss. Additionally, we still employ the fusion of Dice Loss and cross-entropy loss to achieve label fusion. The formula can be defined as:

$$L_{GSCA} = L_{residual} + \gamma \left(\frac{1}{C} \sum_{i=1}^{C} (1 + \phi) \odot -\frac{1}{C} \sum_{i=1}^{2} \log(p_i) B_i \right) \quad (2)$$

B_i means the boundary ground truth of kidney, tumor and cyst. γ is a learnable factor and C equals 4. ϕ is the generated maps that were used to penalize prediction errors.

2.5   Training and Validation Strategies

During the training phase, we utilized a base experimental environment with an RTX 2080 Ti GPU having 12 GB of memory and Python 3.6. Consequently, our maximum batch size was set to 2. We followed the configuration of nnUNet for the learning rate and optimizer. Each step of the network was trained for a total of 1000 epochs. Due to the time-consuming nature of the training process, we randomly selected training and testing sets without performing 5-fold cross-validation.

2.6   Ensembling and Post-processing Method

In our approach, we utilize the results of kidney segmentation from the fine segmentation by 3D residual U-Net as the final kidney segmentation. Then, we incorporate the results of tumor and cyst segmentation from the GSCA-Net into the kidney segmentation region. If the identified tumor or cyst regions are located outside the boundaries of the kidneys, they are directly discarded.

3   Results

3.1   Metric

Our method employs the official evaluation metrics from KiTS2023, which include the Sørensen-Dice score and surface Dice. The Dice coefficient is the most commonly used evaluation metric in the field of image segmentation. Sørensen-Dice can be write as:

$$Dice = \frac{2TP}{FP + FN + 2TP} \quad (3)$$

where TP represents True Positive regions, FP represents False Positive regions, and FN represents False Negative regions. The surface Dice measures the difference on the surface of the segmented regions and the ground truth.

3.2 Experiment Results

Table 1. Ablation study on the KiTS2023 for GSCA-Net.

Network	Sørensen-Dice			Surface Dice		
	kidney	masses	tumor	kidney	masses	tumor
3D U-Net	0.9079	0.6348	0.5488	0.8454	0.4888	0.3997
3D residual U-Net(baseline)	0.9080	0.6367	0.5634	0.8503	0.4940	0.4174
GSCA-Net	0.9605	0.7755	0.7126	0.8734	0.5816	0.5108
GSCA-Net+CBAM	0.9591	0.7680	0.6997	0.8604	0.5676	0.4874
GSCA-Net+Boundary Loss	**0.9606**	**0.7756**	**0.7189**	**0.9056**	**0.6468**	**0.5874**

In this study, we present an improved segmentation model for kidney tumor segmentation by incorporating a boundary loss function and evaluating the impact of different attention mechanisms. As shown in Table 1, Our experiments were conducted on the local test set including cases 500–588. In GSCA-Net, We achieved remarkable results, with Sørensen-Dice coefficients of 0.9591, 0.7755, and 0.7126 for the kidney, masses, and tumor respectively. Surprisingly, the addition of the CBAM (Convolutional Block Attention Module) [14] for local attention led to a reduction in our test results. The inclusion of the boundary loss function yielded significant improvements in Surface Dice, elevating them to 0.9506, 0.6468, and 0.5874 for the kidney, masses, and tumor respectively. As shown in Fig. 5, we presented the segmentation results of some CT image slices, where the red color represents the kidney region, yellow represents the tumor region, and blue represents the cyst region. We observed that GSCA-Net achieves precise segmentation of tumor and cyst shapes.

We have redefined the test sets, with cases 0–99 now being allocated to the test set and case 100–588 as training and validation set. The experimental results obtained are as follows: Sørensen-Dice coefficients of 0.9601, 0.8408, and 0.8466, respectively, for the kidney, tumor, and cyst. Additionally, the Surface Dice coefficients are 0.9273, 0.7298, and 0.7315 for the kidney, tumor, and cyst, respectively.

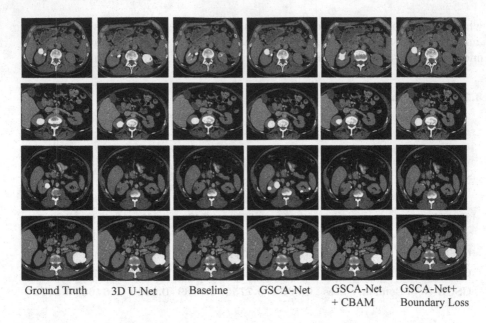

| Ground Truth | 3D U-Net | Baseline | GSCA-Net | GSCA-Net + CBAM | GSCA-Net+ Boundary Loss |

Fig. 5. Examples of segmentation results. Each row from top to bottom represents the sample cases of patients: case580, case588, case503, and case510.

Table 2. Final resoult of GSCA-net on official test set.

Network	Rank	Sørensen-Dice			Surface Dice		
		kidney	masses	tumor	kidney	masses	tumor
GSCA-Net+Boundary Loss	12.5	0.933	0.744	0.679	0.893	0.598	0.534

The results of our network on the official test set are presented in Table 2. Our overall ranking is 12th, with a ranking of 11th for surface Dice and 14th for Dice coefficient. This discrepancy arises due to the implementation of the Boundary Loss, which effectively improves the boundary optimization and results in an increase in Surface Dice. However, it also leads to a decrease in the Dice coefficient.

Since there is no official test set label, we adopt two different networks that have been trained before to show the official test set segmentation results. The segmentation results of GSCA-Net and 3D residual U-Net are compared in Fig. 6. After comparison, our GSCA-Net network is more prone to identifying abnormal areas as cysts (cases 665 and 666), and the identified mass region is larger (cases 638 and 684). In Case 647, the region of renal pelvis dilation or hematoma may have been identified as cyst and tumor region. Our GSCA-Net also tends to classify the same abnormal region as a coexistence of tumor and cyst (case 612 and 665), although tumors and cysts are rarely found in the same connected area.

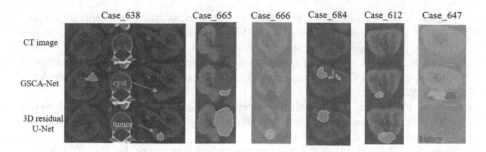

Fig. 6. Special examples of official test set results. The red region represents cysts, the yellow region represents tumors, and the green region represents the kidney. (Color figure online)

4   Discussion and Conclusion

In this paper, we use 3D residual U-Net for ROI extraction and fine kidney segmentation. Then, we propose a GSCA-Net for fine segmentation of tumors and cysts. We also employ 3D residual block as our main block in GSCA-Net, with an increasing number of residual modules in the encoder and two residual blocks in each decoder layer for information extraction. We utilize GSCA module to enhance information extraction capabilities. Finally, we employ a boundary loss to improve Surface Dice results.

Based on the analysis of the official test set, I believe that our method primarily lacks in the classification of tumor or cyst regions. I think that by incorporating an object detection network after the segmentation network, we can assign all segmented connected regions to the same class (tumor or cyst), which should lead to an improvement in the segmentation results.

In the KiTS2023 competition, we would like to provide some suggestions. The extraction of ROIs is time-consuming for participants and does not contribute significantly to innovation. We suggest that the competition organizers provide ROIs. This would allowing participants to allocate more time and effort to focus on improving segmentation results with in ROIs. In addition, after KiTS competition, it would be beneficial to provide individual rankings for cases with overall poorer results. This would allow participants to become familiar with the best outcomes for these special cases, enabling them to better design their own methods to tackle these challenging cases.

Acknowledgment. This research was supported in by the Shandong Natural Science Foundation under Grant No. ZR2019MF003.

References

1. Sung, H., et al.: Global cancer statistics 2020: GLOBOCAN estimates of incidence and mortality worldwide for 36 cancers in 185 countries. CA: A Cancer J. Clin. **71**(3), 209–249 (2021)

2. Merino, M.J., Torres-Cabala, C., Pinto, P., Linehan, W.M.: The morphologic spectrum of kidney tumors in hereditary leiomyomatosis and renal cell carcinoma (HLRCC) syndrome. Am. J. Surg. Pathol. **31**(10), 1578–1585 (2007)
3. Heller, N., et al.: The state of the art in kidney and kidney tumor segmentation in contrast-enhanced CT imaging: Results of the kits19 challenge. Med. Image Anal. **67**, 10182101821 (2021)
4. Isensee, F., Jaeger, P.F., Kohl, S.A., Petersen, J., Maier-Hein, K.H.: nnU-Net: a self-configuring method for deep learning-based biomedical image segmentation. Nat. Methods **18**(2), 203–211 (2021)
5. Zhao, Z., Chen, H., Wang, L.: A coarse-to-fine framework for the 2021 kidney and kidney tumor segmentation challenge. In: Heller, N., Isensee, F., Trofimova, D., Tejpaul, R., Papanikolopoulos, N., Weight, C. (eds.) KiTS 2021. LNCS, vol. 13168, pp. 53–58. Springer, Cham (2022). https://doi.org/10.1007/978-3-030-98385-7_8
6. Golts, A., Khapun, D., Shats, D., Shoshan, Y., Gilboa-Solomon, F.: An ensemble of 3D U-Net based models for segmentation of kidney and masses in CT scans. In: Heller, N., Isensee, F., Trofimova, D., Tejpaul, R., Papanikolopoulos, N., Weight, C. (eds.) KiTS 2021. LNCS, vol. 13168, pp. 103–115. Springer, Cham (2022). https://doi.org/10.1007/978-3-030-98385-7_14
7. George, Y.: A coarse-to-fine 3D U-Net network for semantic segmentation of kidney CT scans. In: Heller, N., Isensee, F., Trofimova, D., Tejpaul, R., Papanikolopoulos, N., Weight, C. (eds.) KiTS 2021. LNCS, vol. 13168, pp. 137–142. Springer, Cham (2022). https://doi.org/10.1007/978-3-030-98385-7_18
8. Bhalerao, M., Thakur, S.: Brain tumor segmentation based on 3D residual U-Net. In: Crimi, A., Bakas, S. (eds.) BrainLes 2019. LNCS, vol. 11993, pp. 218–225. Springer, Cham (2020). https://doi.org/10.1007/978-3-030-46643-5_21
9. Li, X., et al.: Can: context-assisted full attention network for brain tissue segmentation. Med. Image Anal. **85**, 102710 (2023)
10. Gu, R., et al.: CA-Net: comprehensive attention convolutional neural networks for explainable medical image segmentation. IEEE Trans. Med. Imaging **40**(2), 699–711 (2020)
11. Dong, C., Xu, S., Dai, D., Zhang, Y., Zhang, C., Li, Z.: A novel multi-attention, multi-scale 3D deep network for coronary artery segmentation. Med. Image Anal. **85**, 102745 (2023)
12. Gao, Y., Zhou, M., Metaxas, D.N.: UTNet: a hybrid transformer architecture for medical image segmentation. In: de Bruijne, M., et al. (eds.) MICCAI 2021 Part III 24. LNCS, vol. 12903, pp. 61–71. Springer, Cham (2021). https://doi.org/10.1007/978-3-030-87199-4_6
13. Jin, Y., Han, D., Ko, H.: Trseg: transformer for semantic segmentation. Pattern Recogn. Lett. **148**, 29–35 (2021)
14. Woo, S., Park, J., Lee, J.Y., Kweon, I.S.: Cbam: convolutional block attention module. In: Proceedings of the European conference on computer vision (ECCV), pp. 3–19 (2018)

Genetic Algorithm Enhanced nnU-Net
for the MICCAI KiTS23 Challenge

Tao Li[2,3,4], Di Liu[1,2,3], Bo Yang[1(✉)], Yifan Li[1], and Cheng Zhen[1]

[1] AIFUTURE Lab, Beijing 100088, China
yeungbo@gmail.com
[2] National Digital Health Center of China Top Think Tanks, Beijing Normal University,
Beijing 100875, China
[3] School of Journalism and Communication, Beijing Normal University, Beijing 100875, China
[4] China Academy of Social Management, Beijing Normal University, Beijing 100875, China

Abstract. Deep learning-based segmentation techniques have been gaining increasing attention in recent years due to their potential in various medical image segmentation tasks, particularly in the segmentation of kidneys, renal tumors, and renal cysts. One of the major challenges in medical image segmentation is the scarcity of high-quality training data, which often limits the effectiveness and robustness of segmentation algorithms. To address this issue, a novel genetic algorithm (GA) based approach that combines nnU-Net framework is proposed to improve the robustness of medical image segmentation. The proposed approach involves a two-stage process. In the first stage, a set of convolutional neural network (CNN) models are trained with loss function. In the second stage, GA is applied to evolve a population of CNN models with different sets of hyperparameters. This results in a final CNN model with improved robustness and better segmentation performance.

Keywords: genetic algorithm · semantic segmentation · nnU-Net

1 Introduction

Deep learning-based segmentation is a challenging medical image analysis task because medical images are complex and contain multiple features that need to be taken into consideration. To overcome this challenge, researchers have developed various deep learning models such as Convolutional Neural Networks (CNNs) [1], Support Vector Machines (SVMs), and Random Forests (RFs) to segment the images. However, these models are limited by the availability of high-quality training data. In some cases, the training data is insufficient or biased, leading to suboptimal segmentation performance. Therefore, there is a need to develop a robust approach that can handle the limited training data effectively.

© The Author(s), under exclusive license to Springer Nature Switzerland AG 2024
N. Heller et al. (Eds.): KiTS 2023, LNCS 14540, pp. 77–82, 2024.
https://doi.org/10.1007/978-3-031-54806-2_11

2 Methods

In this paper, we proposed a novel GA-based approach that combines Genetic Algorithm [2] with nnU-Net [3] framework to improve the robustness of deep learning-based segmentation systems.

2.1 Training and Validation Data

Our submission made use of the official KiTS23 [4] cohort dataset alone.

The dataset includes patients who underwent cryoablation, partial nephrectomy, or radical nephrectomy for suspected renal malignancy between 2010 and 2022 at an M Health Fairview medical center. Each case's most recent contrast-enhanced preoperative scan (in either corticomedullary or nephrogenic phase) was segmented for each instance of the following three semantic classes. 1.) Kidney: Includes all parenchyma and the non-adipose tissue within the hilum; 2.) Tumor: Masses found on the kidney that were pre-operatively suspected of being malignant; 3.) Cyst: Kidney masses radiologically (or pathologically, if available) determined to be cysts. In modeling, we utilize the pre-released training dataset consisting of 489 cases for the process. As for the competition results, we leverage the remaining 110 cases of validation dataset to generate the prediction results by MICCAI 2023 Leaderboard, as shown in the attached Supplementary Material file.

2.2 Preprocessing

Our submission made use of the official nnU-Net framework preprocessing data format. We download the training dataset and store it in a specific format as nnU-Net requirements. Due to nnU-Net's roots in the Medical Segmentation Decathlon (MSD), its dataset is heavily inspired but has since diverged from the format used in the MSD.

Datasets consist of three components: raw images, corresponding segmentation maps and a dataset.json file specifying some metadata. (More details can be found at [5].)

Specially, in our preprocessing, we split the pre-released training dataset into local training dataset with 410 cases and local test dataset with 79 cases.

2.3 Proposed Method

Our proposed approach is a GA-based method for improving the robustness of deep learning-based nnU-Net segmentation systems. GA-nnUNet consists of two main components: Genetic Algorithm (GA) and nnU-Net. The GA-nnUNet algorithm generates a population of solutions to a given problem by using a set of candidate solutions. nnU-Net is a unifying architecture that is used to integrate different GA population.

In our proposed approach, we first train a set of CNNs using the local trainining dataset alone, which contains 410 images of kidney tumors, renal cysts, and renal lesions. The CNNs are trained using a backpropagation algorithm with a set of different hyperparameters. The CNN architecture consists of an input layer, a convolutional layer, and a pooling layer. The loss function used during training is the binary cross-entropy loss (Fig. 1).

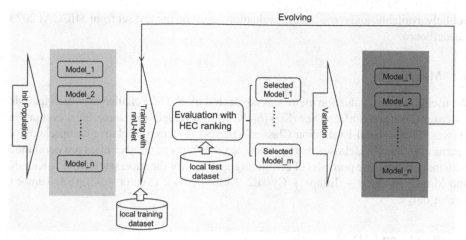

Fig. 1. The brief introduction of GA-nnUNet approach. There are n parallel models are trained on local training dataset, and evaluated on local test dataset for HEC ranking. The top m models are selected for variation to generate next population.

The GA components including gene encoding of hyperparameters (parameters in input layer, convolutional layer size and pooling layer size) for genetic operations (selection, crossover and mutation), and fitness function using "Hierarchical Evaluation Classes" (HECs) ranking score.

After training the CNN models, we apply GA to evolve a population of CNN models with different sets of gene encoded hyperparameters encoded within gene. The GA components used in our approach are two-stage processes: the selection and variation. The GA component selection process selects a subset of parent population to be used in the next generation. The GA component variation process applies different selection rules and variation strategies to the selected individuals.

The selection rules and variation strategies used in our approach are based on a genetic algorithm optimization technique. We use the selection function to evaluate the fitness of the CNN models and select the best candidates for the GA. We use the variation function to combine different hyperparameters and create new population.

Finally, we apply GA to evolve a final CNN model with improved robustness and better segmentation performance.

3   Results

3.1   Dataset

We evaluated the proposed method on the KiTS23 dataset. The KiTS23 dataset includes patients who underwent partial or radical nephrectomy for suspected renal malignancy between 2010 and 2022 at an M Health Fairview medical center. KiTS23 dataset is composed of 599 cases with 489 allocated to the training set and 110 in the test set. Many of these in the training set were used in previous challenges. Note that we first perform the evaluation of our method on the training set because the test set is not publicly available, and retrieved the evaluation result on the test set from MICCAI 2023 Leaderboard.

3.2 Metrics

We used the same evaluation metrics as advocated by KiTS23 challenge, which include Sørensen-Dice and Surface Dice (SD) [6]. KiTS23 leverages the hierarchical evaluation classes "Hierarchical Evaluation Classes" (HECs) to obtain a relative comprehensive measure. In an HEC, classes that are considered subsets of another class are combined with that class for the purposes of computing a metric for the superset. HECs: 1. Kidney and Masses:(Kidney + Tumor + Cyst) 2. Kidney Mass: (Tumor + Cyst) 3. Tumor:(Tumor only).

3.3 Results on KiTS23

We reported the results on KiTS23 challenge training set through selected fold validation and five-fold cross-validation. All the models are evolved with the fitness function of HECs presented in Sect. 3.2 (Fig. 2).

Table 1. Top 8 models of final population of evovling that every model is trained 200 epochs in each generation, all model evaluated on the local test dataset is selected from pre-released training dataset. The simple-net is the model evolved with simpler pool_op_kernel_sizes & conv_kernel_sizes than default setting, while the complex-net is the one evolved with more complex pool_op_kernel_sizes & conv_kernel_sizes.

Evolved Model	Local Rank	Mean_Dice	Mean_SD	Tumor_Dice
3d_lowres_simple-net_on_fold3	1	0.826	0.741	0.73
3d_lowres_simple-net_on_fold0	2	0.822	0.733	0.73
3d_lowres_on_five-fold	3	0.821	0.729	0.728
3d_lowres_complex-net_on_fold3	4	0.804	0.702	0.699
3d_lowres_simple-net2_on_fold0	5	0.749	0.635	0.635
2d_simple-net_on_fold1	6	0.736	0.619	0.603
2d_on_five-fold	7	0.712	0.6	0.565
2d_w_lr5-e5_on_five-fold	8	0.704	0.591	0.552

Based on local ranking in Table 1, we noticed that model with simpler architecture-based parameters shows better performance, and some special training dataset split (such as fold3 in the case) can help to improve the efficiency of training with fewer cross-validation (all of the populations training work done within 2 weeks on single NVIDIA A40 GPU, and the project took 1 month in total.). Finally, we selected the model "3d_lowres_simple-net_on_fold3" to submit evaluation on KiTS23 test dataset.

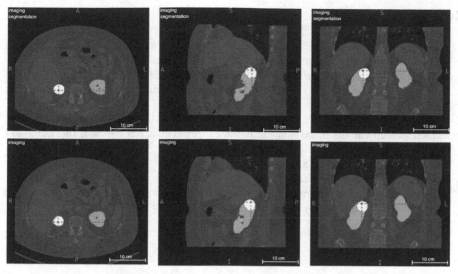

Fig. 2. Examples of predictions next to human-labels on case_00588 used in the local test set. The above is 3D ground truths labeled by human, below is the predictions with our final model.

4 Discussion and Conclusion

This paper presents a novel approach, referred to as GA-nnUNet, to tackle kidney and tumor segmentation tasks. Existing nnU-Net frameworks are found to have limited further improvements, and often possess extensive architecture-based parameters, making them difficult to optimize. GA-nnUNet, on the other hand, employs a refined genetic algorithm (GA) to evolve a model that outperforms mainstream methods in the segmentation of kidneys, renal tumors, and renal cysts. The GA-nnUNet framework utilizes a condensed but flexible search space to restrict the number of architecture-based parameters within a small range, resulting in a significant reduction in the number of parameters and computational cost. To accelerate the convergence process, the fitness function used is Heaps' Correlated Gaussian Filtering (HECs). The experimental results demonstrate the effectiveness of the proposed approach in achieving improved performance compared to existing methods.

Despite the explicit modeling of global information by the GA, it necessitates a substantial amount of memory due to the parallel evolution of CNN models compared to the nnU-Net process. Our future research endeavors will focus on reducing memory consumption while developing a more efficient and accurate segmentation framework.

Furthermore, by analyzing the evolved results, we discovered that the utilization of various effective architecture and parameter patterns in the model's building blocks significantly enhanced the performance of the KiTS23 segmentation task, which provides instrumental domain knowledge for use in future studies.

Acknowledgements. We would like to express our gratitude to the KiTS2023 organizers and the nnU-Net team. We also want to say thanks to Nicholas Heller for his kind help.

References

1. Albawi, S., Mohammed, T.A., Al-Zawi, S.: Understanding of a convolutional neural network. In: 2017 International Conference on Engineering and Technology (ICET), Antalya, Turkey, pp. 1–6 (2017). https://doi.org/10.1109/ICEngTechnol.2017.8308186
2. Goldberg, D.E., Holland, J.H.: Genetic algorithms and machine learning. Mach. Learn. **3**(2), 95–99 (1988)
3. Isensee, F., Jaeger, P.F., Kohl, S.A., Petersen, J., Maier-Hein, K.H.: NnU-Net: a self-configuring method for deep learning-based biomedical image segmentation. Nat. Methods **18**(2), 203–211 (2021)
4. https://kits-challenge.org/KiTS23/
5. https://github.com/MIC-DKFZ/nnUNet
6. Stanislav, N., et al.: Deep learning to achieve clinically applicable segmentation of head and neck anatomy for radiotherapy. arXiv preprint arXiv:1809.04430 (2018)

Two-Stage Segmentation Framework with Parallel Decoders for the Kidney and Kidney Tumor Segmentation

Zhengyu Li[1], Yanjun Peng[1,2](✉), and Zengmin Zhang[1]

[1] College of Computer Science and Engineering, Shandong University of Science and Technology, No. 579, Qianwan'gang Road, Qingdao 266590, China
pengyanjuncn@163.com
[2] Shandong Province Key Laboratory of Wisdom Mining Information Technology, No. 579, Qianwan'gang Road, Qingdao 266590, China

Abstract. Kidney cancer is one of the most common cancers worldwide. Automated segmentation of kidneys, tumors, and cysts from CT images is an important pathway to assist doctors in diagnosis. However, the diverse morphologies of tumors and cysts pose challenges regarding their difficult identification and unpredictable behavior. This paper proposes a two-stage segmentation method, proceeding from coarse to fine. In the first stage, we obtain the kidney region of interest (ROI) based on nnU-Net as input for the second stage. In the second stage, we design a parallel encoder structure. It employs a dual-stream end-to-end training approach, simultaneously monitoring and segmenting boundary information and targets. In particular, a residual channel attention mechanism was incorporated with the boundary prediction branch, highlighting the most relevant feature channels. The method has been experimentally demonstrated to be significantly superior to the baseline nnU-Net. On the official test set, our Kidney + Masses Dice and Tumor Dice are 0.936 and 0.670, respectively, ranking 14th on the leaderboard.

Keywords: Kidney segmentation · Coarse-to-fine framework · Channel attention

1 Introduction

Kidney cancer is one of the top 13 common cancers worldwide. More than 330,000 new cases are diagnosed each year, and its incidence continues to show an increasing trend [1]. Kidney masses, including tumors and cysts, have been the leading cause of Kidney cancer, and precise and quantitative assessment of Kidney masses has become a practical approach for future treatments. Due to the unpredictable shape, the unclear texture and boundaries of the tumor masses within the patient's body, accurately segmenting tumors and cysts from 3D CT images remains a challenging task [2,3]. Semantic segmentation plays a vital role in

© The Author(s), under exclusive license to Springer Nature Switzerland AG 2024
N. Heller et al. (Eds.): KiTS 2023, LNCS 14540, pp. 83–92, 2024.
https://doi.org/10.1007/978-3-031-54806-2_12

aiding diagnosis in the medical field. In this paper, we employ semantic segmentation for Kidney tumor diagnosis, where each voxel in the CT scan is annotated as background, kidney, renal tumor, or cyst.

U-Net [4] is one of the most commonly used convolutional neural network (CNN) architectures for semantic segmentation in biomedical image segmentation applications. It can accurately capture feature information at different scales and, through fusion operations, combine the low-level semantic features from the encoder with the high-level semantic features from the decoder, thereby enhancing segmentation accuracy. A variant of U-Net, nnU-Net [5], demonstrated superior segmentation performance in the 2019 KiTS competition. In the 2021 KiTS Challenge, Zhao et al. [6] proposed a coarse-to-fine segmentation structure based on the nnU-Net as the underlying segmentation network and achieved first place in the competition. George et al. [7] proposed a cascaded U-Net segmentation network from coarse to fine. A two-stage approach is implemented, where the ROI regions are extracted first, and the second stage is trained using the ROI regions. Zhou et al. [8] created a boundary prediction network to address the problem of ambiguous tumor boundaries by generating boundary-aware features and using the boundary information for accurate tumor segmentation.

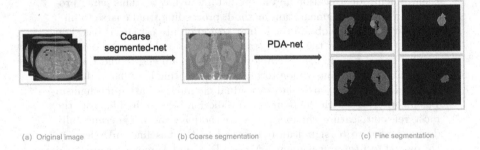

(a) Original image   (b) Coarse segmentation   (c) Fine segmentation

Fig. 1. Flowchart of the two-stage segmentation method from coarse to fine

However, the above methods focus on segmenting the tumor of the whole region and ignore some valuable boundary information. To address the above problems, some works have introduced region boundary constraints [9] or constructed a multitasking framework to extract contour information [10] to improve the segmentation performance. Following the coarse-to-fine segmentation framework, Zhou et al. [11] created a boundary prediction network to address the problem of ambiguous tumor boundaries by generating boundary-aware features and using the boundary information for accurate tumor segmentation. Furthermore, in order to improve the segmentation of foreground regions, an attention mechanism was introduced into the original U-Net. Attention mechanisms have been shown to be effective in natural language processing and computer vision. Several variants based on the fusion attention mechanism of U-Net have been proposed in medical image segmentation tasks [12–14]. The attention mechanism guides the network to focus on essential features, improving parameter efficiency and segmentation accuracy. A residual channel attention mechanism is added for the jump connections of the boundary extraction network. The boundary segmentation is refined by using more boundary information retained in the low-level features to supplement the boundary information extracted in the upsampling process. In this paper, we propose a dual decoder structure that adds a residual channel attention mechanism to the jump connections of the boundary network to capture more boundary information and better assist in target segmentation.

2   Methods

The two-stage segmentation method from coarse to fine is shown in Fig. 1. In the first stage, the raw image is subjected to coarse segmentation to extract a partial CT image containing only the kidney region, i.e., extracting the Region of Interest (ROI). In the second stage, the segmentation network focuses only on the area of interest (ROI) and performs fine segmentation to generate a detailed segmentation map. Specifically, we employed an extended nn-UNet network in the first stage for coarse segmentation, and in the second stage, we used PDA-net for fine segmentation. The network architecture of the first stage is shown in Fig. 2, and the network architecture of the second stage is shown in Fig. 3.

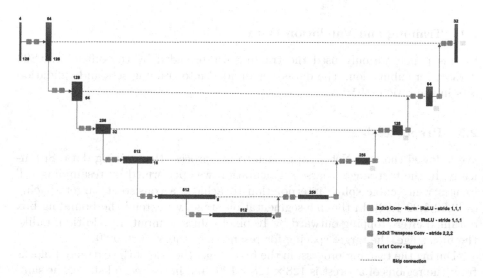

Fig. 2. Coarse segmentation framework employs the extending nnUNet.

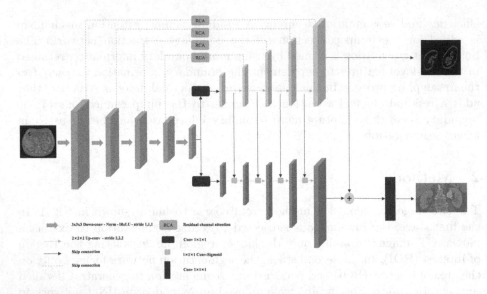

Fig. 3. Parallel-decoder residual channel attention network architecture. A channel attention module was added at the skip connections of the boundary prediction network, highlighting important feature channels of the boundaries, enriching boundary information, and better assisting the overall segmentation objective.

2.1   Training and Validation Data

In this paper, we only used the training set provided by the official KiTS23 dataset for submission. The dataset is divided into training sets and validation sets in the ratio of 4:1.

2.2   Preprocessing

We followed the nnU-Net [5] approach to pre-process the training data. Specifically, in the first stage, coarse segmentation was performed by resampling CT images using cubic spline interpolation to achieve a consistent target spacing of [3.22 1.92 1.92]. In the fine segmentation stage, we extend the bounding box obtained after cropping outward by 10-pixel values as input. In addition, unlike the first stage, the target spacing for resampling is [0.80 0.80 0.80].

During the training process, in the first stage, the size of the extracted blocks from the regions of interest is $128 \times 128 \times 128$, and in the second stage, the sampled block size is $128 \times 224 \times 85$. Subsequently, intensity clipping was performed using the default settings of nnU-net (0.5 and 99.5 percentiles) with a clipping range of $[-90, 405]$ Hounsfield Units (HU). We subtract the mean value of 101.9 and divide by the standard deviation of 72.63 to perform normalization on the input images, data augmentation techniques such as rotation, scaling, intensity transformations, and others were applied.

2.3 Proposed Method

Coarse Segmentation. Through experiments, we have found that cropping out the portion of the CT image containing only the kidney region from the original complete CT image can improve the segmentation results of the kidneys and renal masses (tumors and cysts) to a certain extent. Therefore, we first utilize the extending nnU-Net [15] to obtain kidney ROIs from CT images of each case, aiming for more precise segmentation results.

Fine Segmentation. Analyzing the provided dataset, we discovered valuable boundary information that should be addressed. Therefore, our dual-decoder architecture consists of boundary prediction and object segmentation branches. We use the kidney ROI as input and segment the boundary mask, kidney mask, tumor mask, and cyst mask separately on the fine segmentation network with dual decoders. The encoder part adopts the concept of extending nnU-Net, increasing the number of channels, and includes five levels of convolutional layers with the same resolution. Each layer consists of a 3D convolution with a $3 \times 3 \times 3$ kernel and strides of 1 in each dimension. Expressly, a residual channel attention mechanism is incorporated at the skip connections of the boundary prediction network, which enhances the extraction and supplementation of boundary information. Additionally, we utilize residual blocks instead of regular convolutional blocks in the convolutional layers of the target segmentation branch.

Residual Channel Attention Module. Channel attention has effectively utilized the most informative feature channels while suppressing irrelevant ones, automatically identifying and emphasizing the relevant feature channels [16]. Therefore, we introduced residual channel attention at the skip connections in our fine segmentation network to align the low-level features from the encoder with the high-level features from the decoder. The feature channels from the encoder contain richer low-level information, while the feature channels from the

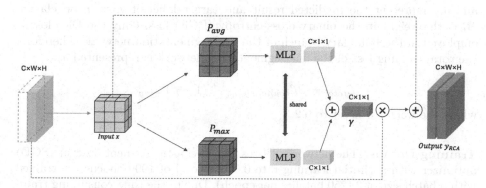

Fig. 4. Structure of our residual channel attention module with the residual connection.

decoder contain more high-level information. We employ this attention mechanism to compensate for the lost low-level information during the upsampling process, thus complementing the boundary information. As shown in Fig. 3.

We assume X represents the concatenated input feature maps with C channels. First, global average pooling and global maximum pooling are utilized to obtain global information for each channel, and then output and represent it as $p_{avg}(x) \in R^{C \times 1 \times 1}$ and . To obtain the channel coefficient $\gamma \in [0,1]^{C \times 1 \times 1}$ using Multi-Layer Perceptron (MLP) M^r, and M^r consists of two fully connected layers, with the first layer having an output channel number of C/r, followed by ReLU, and the second layer having an output channel number of C. $p_{avg}(x)$ and $p_{max}(x)$ share the use of M^r, and their results are summed and fed into a Sigmoid function to obtain γ. Our residual channel attention module's output is: $y_{RCA} = x \cdot \gamma + x$. We added residual channel attention modules at the skip connections of the network, as shown in Fig. 4.

Loss Function. During the training process, the target detection branches of our coarse and fine segmentation models were trained using a combination of Dice loss and cross-entropy loss. The following equation represents the loss function:

$$\mathcal{L}_{seg} = \alpha \mathcal{L}_{dice} + (1 - \alpha) \mathcal{L}_{BCE} \tag{1}$$

The weight α was set to 0.5 in the experiments. The detailed formulas for the Dice loss and cross-entropy loss are as follows:

$$\mathcal{L}_{dice} = 1 - \frac{2}{3} * \frac{\sum_{i=1}^{3} \sum_{n=1}^{N} (p_{i_n} * q_{i_n})}{\sum_{i=1}^{3} \sum_{n=1}^{N} (p_{i_n} + q_{i_n})} \tag{2}$$

$$\mathcal{L}_{BCE} = \sum_{i=0}^{3} Bi \sum_{n=1}^{N} (q_{i_n} \log(p_{i_n}) + ((1 - p_{i_n}) \log(1 - q_{i_n}))) \tag{3}$$

N represents the voxel number, and i represents the index of each voxel. p_{i_n} and q_{i_n} represent the predicted result and target label of voxel n on class i, Bi is the weight in the binary cross-entropy (BCE) loss. Only the Dice loss is employed in the boundary branch of the fine segmentation network. Therefore, the total training loss of the fine segmentation network is represented as:

$$\mathcal{L}_{total} = \beta \mathcal{L}_{boundary\,dice} + (1 - \beta) \mathcal{L}_{seg} \tag{4}$$

where the weight β is set to 0.2.

Training Details. The network utilizes the stochastic gradient descent (SGD) optimizer with an initial learning rate 0.01. A total of 1000 epochs are trained with a batch size of 2 (250 batches per epoch). Due to the time-consuming training process, we did not employ the 5-fold cross-validation in nnU-Net. Instead, we used the first 398 training data as the training set, while the rest were allocated as the validation set. The remaining hyperparameters mainly follow the default values of nnU-Net. We implemented our network using PyTorch on a single NVIDIA GeForce RTX 3090 GPU with 24 GB of memory.

3   Results

We summarize all our experimental results in Table 1. All results are based on the validation set of the last 100 cases. The Dice coefficients for the kidneys, renal masses, and renal tumors are 0.96056, 0.755431, and 0.677995, respectively; the average Dice coefficient is 0.8037. The surface Dice scores for the kidney, renal masses, and renal tumors were 0.9054, 0.6304, and 0.5674, respectively. The average surface Dice score is 0.7010. Our method outperforms the baseline nnU-Net significantly in the segmentation of masses and tumors, while there is a slight improvement in the segmentation of the kidneys. Additionally, we selected two variants of nnU-Net as comparative experiments, and the results demonstrated that, under the same experimental conditions, our network model outperforms their models. We selected four representative cases and visualized the segmentation results of all models on them, as shown in Fig. 5.

In Table 2, we showcase the results on the official test set. Based on the visualized results of the predicted labels, the analysis primarily attributes to the incompleteness in tumor recognition, particularly in cases where tumors and cysts overlap. These instances only identify the boundaries of the tumor as well as some scattered portions, failing to recognize the entirety of the tumor. As a result, this leads to a lower Dice score for tumor segmentation.

Table 1. Dice score and Surface Dice of the proposed method and other baselines methods on the validation set.

Model	Dice				Surface Dice			
	Kidney	Mass	Tumor	Ave	Kidney	Mass	Tumor	Ave
Baseline	0.9525	0.7058	0.6179	0.7587	0.8996	0.5832	0.4615	0.6481
3D Extending U-Net	0.9467	0.7124	0.6265	0.7618	0.8982	0.5960	0.5084	0.6675
3D Large U-Net	0.9593	0.7494	0.6745	0.7944	0.9017	0.6107	0.5330	0.6818
Ours	**0.9612**	**0.7554**	**0.6953**	**0.8037**	**0.9054**	**0.6304**	**0.5674**	**0.7010**

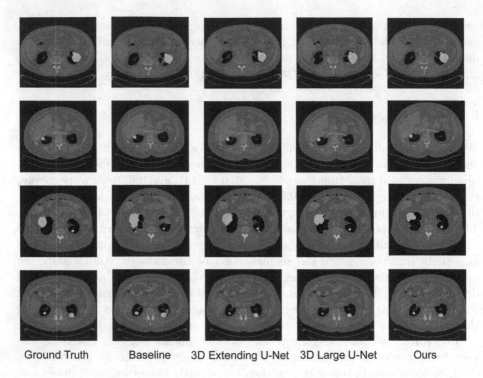

Ground Truth   Baseline   3D Extending U-Net   3D Large U-Net   Ours

Fig. 5. We sequentially selected four different cases and visualized their segmentation results on different models. Red represents the kidneys, yellow represents tumors, and green represents cysts. (Color figure online)

Table 2. Dice score and Surface Dice of the proposed method on the official test set.

Model	Dice				Surface Dice			
	Kidney	Mass	Tumor	Ave	Kidney	Mass	Tumor	Ave
Ours	0.936	0.751	0.670	0.786	0.888	0.599	0.527	0.671

4 Discussion and Conclusion

This paper presents a two-stage segmentation framework, from coarse to fine, for segmenting the kidneys, tumors, and cysts from CT images. nnU-Net serves as the baseline for our model, and we improve it by employing ROI cropping, adding residual channel attention blocks at skip connections, and utilizing an end-to-end training approach with dual-stream. We crop the original images to a suitable size for model training and provide more supervision and supplementation for valuable boundary information, thus aiding in target segmentation. The obtained experimental results are significantly superior to the baseline and the other two variants of nnUNet.

We performed visualizations on the predicted results and observed that in certain samples, the model tends to confuse tumor and cyst identification. In most cases, the tumor recognition region was either too small or misidentified as a cyst. This discovery provides significant assistance for the improvement of our subsequent models. I believe we should not only focus on extracting local information but also consider the integration of global information. Perhaps, we can start with classification and use the classification results to aid in segmentation, emphasizing the importance of a holistic understanding of the target. Furthermore, I believe we can enhance the model performance by improving the loss function. I believe our model has room for improvement, and we will continue optimizing it to contribute to the automated segmentation of renal cancer.

Acknowledgment. This work was supported in part by the National Natural Science Foundation of China (Grant No. 61976126). We want to express our gratitude to the organizers of KiTS2023 and the nnU-Net team for their assistance.

References

1. Chow, W.H., Dong, L.M., Devesa, S.S.: Epidemiology and risk factors for kidney cancer. Nat. Rev. Urol. **7**(5), 245–257 (2010)
2. Li, X., Liu, L., Heng, P.A.: H-Denseunet for kidney and tumor segmentation from CT scans (2019)
3. Yu, Q., Shi, Y., Sun, J., Gao, Y., Zhu, J., Dai, Y.: Crossbar-net: a novel convolutional neural network for kidney tumor segmentation in CT images. IEEE Trans. Image Process. **28**(8), 4060–4074 (2019)
4. Ronneberger, O., Fischer, P., Brox, T.: U-Net: convolutional networks for biomedical image segmentation. In: Navab, N., Hornegger, J., Wells, W.M., Frangi, A.F. (eds.) MICCAI 2015. LNCS, vol. 9351, pp. 234–241. Springer, Cham (2015). https://doi.org/10.1007/978-3-319-24574-4_28
5. Isensee, F., Jaeger, P.F., Kohl, S.A., Petersen, J., Maier-Hein, K.H.: nnu-net: a self-configuring method for deep learning-based biomedical image segmentation. Nat. Methods **18**(2), 203–211 (2021)
6. Zhao, Z., Chen, H., Wang, L.: A coarse-to-fine framework for the 2021 kidney and kidney tumor segmentation challenge. In: Heller, N., Isensee, F., Trofimova, D., Tejpaul, R., Papanikolopoulos, N., Weight, C. (eds.) KiTS 2021. LNCS, vol. 13168, pp. 53–58. Springer, Cham (2022). https://doi.org/10.1007/978-3-030-98385-7_8
7. George, Y.: A coarse-to-fine 3D U-Net network for semantic segmentation of kidney CT scans. In: Heller, N., Isensee, F., Trofimova, D., Tejpaul, R., Papanikolopoulos, N., Weight, C. (eds.) KiTS 2021. LNCS, vol. 13168, pp. 137–142. Springer, Cham (2022). https://doi.org/10.1007/978-3-030-98385-7_18
8. Zhao, Z., Chen, H., Li, J., Wang, L.: Boundary attention u-net for kidney and kidney tumor segmentation. In: 2022 44th Annual International Conference of the IEEE Engineering in Medicine & Biology Society (EMBC), pp. 1540–1543. IEEE (2022)
9. Fang, Y., Chen, C., Yuan, Y., Tong, K.: Selective feature aggregation network with area-boundary constraints for polyp segmentation. In: Shen, D., et al. (eds.) MICCAI 2019. LNCS, vol. 11764, pp. 302–310. Springer, Cham (2019). https://doi.org/10.1007/978-3-030-32239-7_34

10. Murugesan, B., Sarveswaran, K., Shankaranarayana, S.M., Ram, K., Joseph, J., Sivaprakasam, M.: Psi-net: shape and boundary aware joint multi-task deep network for medical image segmentation. In: 2019 41st Annual International Conference of the IEEE Engineering in Medicine and Biology Society (EMBC), pp. 7223–7226. IEEE (2019)
11. Zhou, T., et al.: Cross-level feature aggregation network for polyp segmentation. Pattern Recogn. **140**, 109555 (2023)
12. Jin, Q., Meng, Z., Sun, C., Cui, H., Su, R.: Ra-UNet: A hybrid deep attention-aware network to extract liver and tumor in CT scans. Front. Bioeng. Biotechnol. **8**, 605132 (2020)
13. Oktay, O., et al.: Attention u-net: learning where to look for the pancreas. arXiv preprint arXiv:1804.03999 (2018)
14. Li, C., et al.: Attention unet++: a nested attention-aware u-net for liver CT image segmentation. In: 2020 IEEE International Conference on Image Processing (ICIP), pp. 345–349. IEEE (2020)
15. Luu, H.M., Park, S.H.: Extending nn-UNet for brain tumor segmentation. In: Crimi, A., Bakas, S. (eds.) BrainLes 2021. LNCS, vol. 1263, pp. 173–186. Springer, Cham (2021). https://doi.org/10.1007/978-3-031-09002-8_16
16. Gu, R., et al.: Ca-net: comprehensive attention convolutional neural networks for explainable medical image segmentation. IEEE Trans. Med. Imaging **40**(2), 699–711 (2020)

3d U-Net with ROI Segmentation of Kidneys and Masses in CT Scans

Connor Mitchell[1(✉)], Shuwei Xing[1,2], and Aaron Fenster[1,2,3]

[1] Robarts Research Institute, The University of Western Ontario,
London, ON, Canada
cmitc@uwo.ca
[2] School of Biomedical Engineering, Western University, London, ON, Canada
[3] Department of Medical Biophysics, Western University, London, ON, Canada

Abstract. This project focuses on automatic kidney, tumor and cyst segmentation to assist doctors in diagnosing kidney cancer. We created a deep learning model using methods to first isolate the region of interest of the kidneys, then to segment the kidney and its masses. We used the TotalSegmentator tool to obtain a rough segmentation of the kidneys, then during pre-processing, expanded this region of interest by 18 pixels. This new region of interest was inputted into a 3d segmentation network trained using the nnU-Net library to fully segment the kidneys and masses within them. The current model achieved an average DICE score on validation data of 0.95 for kidney segmentations, and around a 0.8 for tumour and cyst segmentations. On the KiTS23 testing data, the model achieved a 0.94 DICE for kidney segmentations and a 0.73 DICE for mass segmentations.

Keywords: Semantic Segmentation · Kidney Cancer · nnU-Net

1 Introduction

In 2023, there will be an estimated 81,800 new cases of invasive kidney cancer in the United States alone, with an estimated 14,890 deaths resulting from kidney cancer [2]. The early diagnosis of kidney cancer is considered to be an effective way to reduce the incidence and deaths. Clinically, doctors also need to differentiate tumour types, such as slow-growing primary tumors or aggressive metastatic tumors, to determine the most effective treatment. Therefore, an accurate and automatic approach to identify the kidney tumour is the current unmet need. In this project, which takes advantage of deep learning techniques, we propose a U-Net-based workflow with Region of Interest (ROI) masking trained on the KiTS23 dataset to perform semantic segmentations on kidneys and any masses within them.

2 Methods

The method used to segment the kidneys and masses can be broken up into two sections: preprocessing, and semantic segmentation using our U-Net-based

© The Author(s), under exclusive license to Springer Nature Switzerland AG 2024
N. Heller et al. (Eds.): KiTS 2023, LNCS 14540, pp. 93–96, 2024.
https://doi.org/10.1007/978-3-031-54806-2_13

segmentation models. The preprocessing of the images consists of first isolating a "region of interest", or a subset of the original 3d image to isolate a smaller region containing both kidneys. The goal of this is to try to isolate the kidneys and kidney masses in hopes that the accuracy of the deep learning model will be improved in comparison to a model that inputs the entire CT scan during training. Then data augmentation was performed on the isolated regions. Next the region of interest data was used to train a 3d convolutional U-Net developed from the nnU-net [1] framework to segment the kidneys and any masses inside them (Fig. 1).

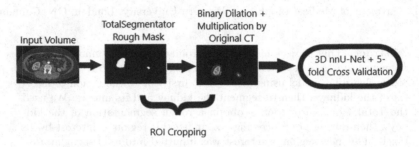

Fig. 1. ROI Segmentation Method

2.1   Training and Validation Data

Our submission made use of the official KiTS23 training set alone.

2.2   Preprocessing

The primary improvement to our method in comparison to a standard U-Net stems from different preprocessing steps. The data was first segmented with the TotalSegmentator tool. TotalSegmentator is a deep learning model trained on 1204 CT examinations used to segment 104 different anatomical structures, including kidneys [3]. It was trained using nnU-Net's 3d_fullres configuration on 4000 epochs. We used TotalSegmentator to obtain a rough, and fairly inaccurate segmentation of the left and right kidneys. Next, we loaded the original CT data and the newly acquired TotalSegmentator masks into Python using SimpleITK's ReadImage function. Binary expansion was performed on each mask to obtain a much broader ROI mask, which was then multiplied by each slice in the original CT volumes using Python's NumPy library. Specifically, this ROI was an expansion of 18 pixels from the original segmentation mask. We used the binary_dilation method, which is a part of SciPy's ndimage module to perform the binary expansion in Python. This removed any unnecessary information from the CT images for the kidney segmentation task. Because of this transform, the HU values for the background space were set to 0, while the HU values within the ROI were left unchanged. Some standard data augmentation was performed on the training data. nnU-Net has some standard preprocessing for the input data, which includes data augmentation. Gaussian blurring, Gaussian noise, rotations, scaling, zooming, and mirroring were all applied.

2.3 Proposed Method

nnU-Net uses a standard U-Net architecture, which was used in both the TotalSegmentator tool during preprocessing, and in the network to train the cropped images on kidney and mass segmentation. Batch normalization is not used in nnU-Net architectures, instead instance normalization is used. DICE loss is used as the metric for the loss function. The network kernel size is 3×3 in every convolutional layer. The dataset of 489 volumes was trained on 5-fold cross validation, each with 1000 epochs. So 5 models trained with 5 different test/validation splits were trained and predictions will be run on an average of the best checkpoints from those 5 models. nnU-Net also makes use of connected component analysis to remove non-connected regions predicted to be a part of the kidney or mass, which are smaller than the largest prediction.

3   Results

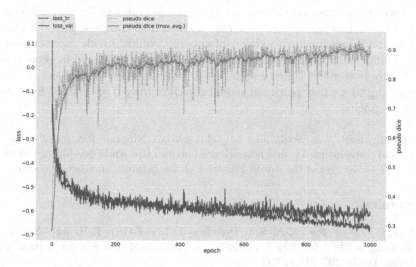

Fig. 2. Change in Pseudo DICE and Loss for Training and Validation Data During Training

The average DICE score on the validation set over the 5 folds of training after 1000 epochs was as follows: for fold 1: 0.76, fold 2: 0.75, fold 3: 0.72, fold 4: 0.73, and fold 5: 0.69. These are aggregated averages over the kidney segmentation, tumor segmentation and cyst segmentations. Additionally, Fig. 2 presents a plot of how the DICE score changes on validation data during the training process for the first fold. We can observe that it as we trained the model for longer the accuracy of the model followed almost a logarithmic curve. It would likely still be beneficial to train the model for longer, if time permitted, in order to get a better overall accuracy. Training all folds took 2 days running on an NVIDIA GRID V100D-32Q. When presented with the KiTS23 test data, our model performed as we expected based on the validation scores. More specifically, we achieved a DICE score of 0.777, with a surface DICE of 0.648. Our DICE score for the kidney and masses together was 0.94, and our DICE score for just the masses (tumor and cysts) was 0.73.

4   Discussion and Conclusion

As the nnU-Net model is already one of the leading models for semantic medical image segmentation, it is very difficult to improve on it. The main limitation to our model was the model architecture. It is possible that using a different type of 3d U-Net, such as the 3d_cascade_fullres architectures, which is another architecture within nnU-Net, would improve the overall performance of the model [4], however the computation cost and training time for this model would be impractical for the period of time we had for building the model.

Previously, our team created a 2d nnU-Net segmentation model using the same workflow as described in this paper. The 2d model's performance was significantly worse than the 3d model's, since it was trained over only 250 epochs in each fold. If time allowed, an ensemble approach to compare the 3d and a 2d model, trained over the same number of epochs, would provide more consistent results. Additionally, training with more epochs for the 3d model in general may yield slightly better DICE loss results. As an extension to this project, it would be interesting to see how much our model could improve if we trained it on more than 1000 epochs.

Acknowledgment. We would like to acknowledge Ningtao Liu for sharing his PyTorch and convolutional neural network wisdom and tips while developing our workflow, as well as the rest of the Aaron Fenster Lab for helping out when necessary.

References

1. Isensee, F., Jaeger, P.F., Kohl, S.A., Petersen, J., Maier-Hein, K.H.: nnU-Net: a self-configuring method for deep learning-based biomedical image segmentation. Nat. Methods **18**(2), 203–211 (2021)
2. Siegel, R.L., Miller, K.D., Wagle, N.S., Jemal, A.: Cancer statistics, 2023. CA Cancer J. Clin. **73**(1), 17–48 (2023)
3. Wasserthal, J., ET AL.: TotalSegmentator: robust segmentation of 104 anatomical structures in CT images. arXiv e-prints arXiv:2208.05868 (2022). https://doi.org/10.48550/arXiv.2208.05868
4. Zettler, N., Mastmeyer, A.: Comparison of 2D vs. 3D U-Net organ segmentation in abdominal 3D CT images. arXiv e-prints arXiv:2107.04062 (Jul 2021). https://doi.org/10.48550/arXiv.2107.04062

Deep Learning-Based Hierarchical Delineation of Kidneys, Tumors, and Cysts in CT Images

Andrew Heschl[1]([✉]), Hosein Beheshtifard[1], Phuong Thao Nguyen[1], Tapotosh Ghosh[1], Katie Ovens[1], and Farhad Maleki[1,2,3]

[1] Computer Science Department, University of Calgary, Calgary, AB, Canada
andrew.heschl@ucalgary.ca
[2] Department of Diagnostic Radiology, McGill University, Montreal, QC, Canada
[3] Department of Radiology, University of Florida, Gainesville, FL, USA

Abstract. Kidney cancer is among the most prevalent forms of cancer. Recent studies have investigated the use of machine learning, especially deep learning, in detecting tumors from kidney CT scans. This paper introduces a hierarchical pipeline for segmenting kidneys, tumors, and cysts in CT images. Our method demonstrates promising outcomes in the segmentation of kidneys, renal cysts, and tumors, showcasing its potential to accurately identify and localize tumors and cysts in CT scans. This can significantly aid early diagnosis and treatment planning.

Keywords: Semantic Segmentation · Medical Imaging · Computer Vision

1 Introduction

Kidney cancer is among the most prevalent forms of cancer. An estimated 81,800 new cases are expected to be diagnosed in the US in 2023 [6], representing only a small fraction of the cases expected worldwide. Accurate diagnosis and timely treatment of kidney cancer are crucial for enhancing patient outcomes and reducing mortality rates. Recent advancements in machine learning, particularly deep learning techniques, have demonstrated tremendous potential in the detection and segmentation of tumors in CT scans [1,2,4,7].

The 2023 Kidney and Kidney Tumor Segmentation Challenge, known as KiTS23, aims to identify the most effective system for the automatic segmentation of kidneys, renal tumors, and renal cysts. This paper represents our submission for the KiTS23 challenge, which was ranked 16th. The primary objective of this research is to develop an accurate and efficient deep-learning solution for kidney tumor segmentation in CT scans. We utilize a hierarchical pipeline that incorporates three models. These models are based on nnU-Net v2 [3]—an automated approach for configuring a U-Net model [5] designed for biomedical image

---

H. Beheshtifard and P.T. Nguyen—These authors contributed equally.

© The Author(s), under exclusive license to Springer Nature Switzerland AG 2024
N. Heller et al. (Eds.): KiTS 2023, LNCS 14540, pp. 97–106, 2024.
https://doi.org/10.1007/978-3-031-54806-2_14

segmentation. By consolidating the predictions from all models, we generate a single, cohesive prediction.

The remainder of this paper is organized as follows: Sect. 2 presents the methodology and the model architecture in detail. In Sect. 3, we present the results, and, finally, in Sect. 4, we discuss the results and conclude the paper.

2   Methods

In the previous KiTS challenge, an analysis of the top-performing submissions revealed that an overwhelming majority of them, specifically 4 out of the top 5, utilized nnU-Net as the foundational framework for their models. This trend serves as strong evidence of the potential of nnU-Net in effectively addressing the challenging task of segmenting kidneys, tumors, and cysts.

The winning submission in the KiTS21 challenge employed a strategy that involved using kidney segmentation as a foundation for predicting tumors and cysts [8]. Recognizing the effectiveness of this approach, we have incorporated a similar logic into our methodology to enhance performance; however, instead of performing a coarse rectangular crop followed by a kidney segmentation, we have devised a hierarchical approach, which is shown in Fig. 1. Initially, model K identifies the kidneys as the region of interest (ROI). Afterward, we perform a post-processing step to remove any object from the segmentation that is not of the desired region. Following this we remove objects which have a voxel count below one-quarter the total count in the predicted mask. We then conduct a dilation operation on the predicted masks using a max-pooling operator with a stride of one and a kernel size of 11. This dilated version is used to mask the raw image, after which we extract each extended kidney region. This produces a new dataset to train our subsequent models for tumor segmentation. Model M predicts the mass region, i.e., tumors and cysts, within each kidney, and then undergoes similar dilation and masking processes. The final model, model T, differentiates tumors from the mass ROIs. We then consolidate these predictions to produce a semantic mask that delineates kidneys, mass, and tumors.

While models M and T operate on relatively small ROIs-cropped versions of the kidneys or mass-model K is applied to the entire image. This poses a challenge, as a typical GPU may not support 3D analysis of full CT images due to its limited VRAM (Video Random Access Memory), even at a batch size of one. To address this, we implemented two strategies. First, we utilized a 3D U-Net architecture with a small patch size of $224 \times 320 \times 256$. Secondly, we developed a 2D U-Net model to segment the kidney on each axial slice of a CT image. Both methods ensure that the analysis can be conducted irrespective of the availability of a high-end GPU.

2.1   Training and Validation Data

Although the use of external datasets for model training was permitted in the KiTS23 challenge, we only utilized the training subset of KiTS23 for model training and validation.

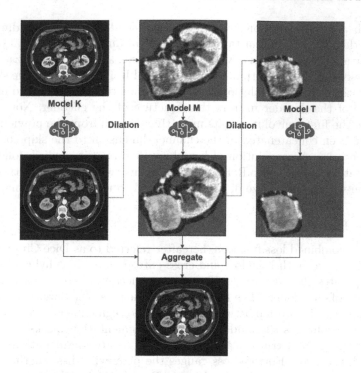

Fig. 1. Our hierarchical pipeline. Kidney prediction is represented in green, mass prediction in yellow, and tumor prediction in pink. (Color figure online)

2.2   Preprocessing

In order to preprocess the CT images, we utilize the preprocessing pipeline provided by nnU-Net. This includes normalization and harmonizing voxel spacing. Normalization is achieved by first clipping the voxel intensity values to fall within the 0.5^{th} and 99.5^{th} percentiles. Subsequently, a conventional z-score normalization is applied.

2.3   Network Architecture

Initially, we adopted nnU-Net v2 as our baseline model to simultaneously segment all regions of interest: kidneys, masses, and tumors. However, we observed a substantial discrepancy in accuracy between kidney segmentation and tumor/mass segmentation. The kidney segmentation exhibited high accuracy, while the segmentation of tumors and masses lagged behind. Recognizing the hierarchical nature of the classes, we sought to enhance our mass and tumor segmentations. As a result, we crafted a pipeline composed of three models, all derived from nnU-Net v2 full-res.

The nnU-Net v2 contains six stages in both encoder and decoder, with skip connections in the form of concatenation at each stage. A stage of the encoder starts with a 3D convolution layer, utilizing a stride of 2 in each dimension. Following this, there is one instance of normalization block and a leaky ReLU, followed by another convolution with a stride of one, another normalization, and another leaky ReLU. The output from this second ReLU is used for the skip connection. Each of these convolutions utilizes a kernel of size 3 in each dimension. Each stage of the decoder mirrors the structure of the encoder. Note that at each stage, the first convolution acts upon the features from the previous stage, which have been concatenated in the channel dimension to the skip connection from the encoder. A convolution of kernel size $3 \times 3 \times 3$ is applied, followed by a normalization and a leaky ReLU. This happens twice. Finally, a transposed convolution with a kernel size of two is employed to up-sample the image.

2.4 Loss Function

We utilize a combined loss function, hereafter referred to as Dice Cross-Entropy loss, which merges both Dice loss and cross-entropy loss. This hybrid loss function incorporates the advantages of these two metrics, thereby optimizing the performance of our models. The Dice loss, also known as the Sørensen-Dice coefficient, evaluates the overlap between the predicted segmentation and the ground truth. This encourages the model to produce segmentation masks that closely align with the ground truth, leading to more accurate segmentations. On the other hand, the Cross-Entropy loss gauges the pixel-wise dissimilarity between the predicted classes and the ground truth. By minimizing this loss, the model is prompted to assign higher probabilities to the correct pixel classes, thereby enhancing the accuracy of the segmentation results.

2.5 Optimization Strategy

In our optimization strategy, we employ Stochastic Gradient Descent (SGD). To further enhance the optimization process, we adopt a polynomial learning rate schedule. The learning rate is a crucial hyperparameter that dictates the step size during parameter updates. The polynomial learning rate schedule typically involves decreasing the learning rate gradually as training progresses. This approach helps the model converge more effectively by initially taking larger steps to explore the parameter space and then gradually reducing the step size to make smaller, fine-tuned adjustments. The schedule is as follows:

$$i \times (1 - \frac{e}{t})^{0.9}$$

where i is the initial learning rate, e is the current epoch, and t is the total number of epochs. For this and the remaining optimizer hyperparameters, we used the default values as suggested by the nnU-Net pipeline.

2.6 Validation Strategy

We trained a semantic segmentation model using a randomly selected validation fold, which constituted one-fifth of the total training dataset. Due to our limited access to computational resources, we opted to not conduct a K-fold cross-validation. While a single train-validation split might not capture the variability of the data as effectively as full cross-validation would, it nonetheless allows us to progress in the development and evaluation of our models for this segmentation task.

2.7   Post-processing

In the post-processing step during inference, we use a method to refine the predicted total region by removing any portions where the predicted volume constitutes less than one quarter of the overall predicted mass. This step helps filter out the small regions of the images where the models may make extraneous predictions.

Additionally, in our hierarchical approach, we apply a dilation operation to the resulting masks to further enhance the downstream predictions. By dilating the mask, we expand the ROI, effectively enlarging the area surrounding the remaining kidney or mass. This approach allows us to improve the prediction on kidneys and focus the tumor or cyst prediction on the pertinent ROIs. By omitting irrelevant regions, we consequently reduce computational costs.

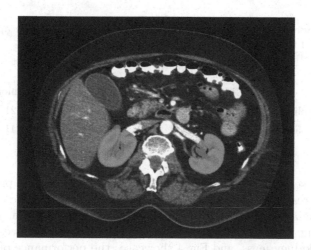

Fig. 2. The prediction of model K for case 97 of KiTS23 dataset compared to the ground truth. The area with red color represents the kidney and masses predicted, and the area with green color represents the ground truth. (Color figure online)

3   Results

As required by the KiTS organizers, we use "Hierarchical Evaluation Classes" (HECs) instead of each ROI alone. The HECs are (1) Kidney and Masses (Kidney + Tumor + Cyst), (2) Kidney Mass (Tumor + Cyst), and (3) Tumor only.

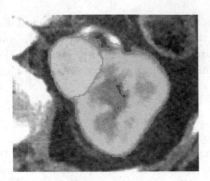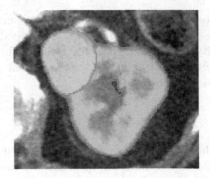

Fig. 3. Left: the ground truth. Right: The prediction of model M for case 10 (red represents kidney and blue represents mass). (Color figure online)

We employ the Sørensen-Dice coefficient and Surface Dice score to assess the efficacy of our approach. We draw comparisons with the standard nnU-Net model—i.e., our baseline—to gain deeper insight into our pipeline's performance. As shown in Table 1, our model consistently improves prediction accuracy for kidneys, masses, and tumors.

Table 1. The Sørensen-Dice coefficient and Surface Dice scores of our approach compared to those of the baseline on our validation set.

	Sørensen-Dice			Surface Dice		
	Kidney	Mass	Tumor	Kidney	Mass	Tumor
Baseline	0.965	0.780	0.741	0.941	0.653	0.621
Our approach	0.971	0.829	0.816	0.949	0.693	0.641

As can be seen from the results, our pipeline works well in predicting the regions of interest. Visualizations of the models' predictions were conducted to understand the areas of struggle. Figure 2 illustrates the performance of model K in predicting kidney and masses, Fig. 3 shows the performance of model M in predicting kidney mass, and Fig. 4 showcases the performance of model T in predicting tumors.

The example in Fig. 2 shows that model K can segment the kidney with high accuracy. However, as shown in Fig. 5, the models tend to have difficulty accurately capturing small parts that branch off distantly from the main area of the kidneys. Figure 3 and Fig. 4 show that our models' predictions of masses and tumors are precise. In case 244, for example, our model seems to generate a better result than the ground truth. However, in some cases, they may struggle to locate and segment small or complex tumors and masses.

Notably, in anomaly cases, for example, in case 194, where there is only one kidney in the image, our models are able to detect it correctly without

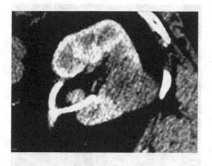
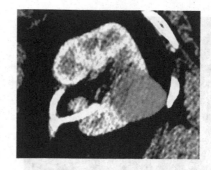

Fig. 4. The prediction from model M for case 244 is depicted in red, in contrast to the ground truth in green. We visualized this case with both outlining (left) and filling (right) the ROIs. (Color figure online)

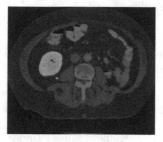

Fig. 5. The prediction of model K (red) for case 194 compared to the ground truth (green). (Color figure online)

segmenting another region as a kidney, as evidenced in Fig. 5. The correct segmentation in anomaly cases like this provides strong evidence that our model is effectively learning the distinguishing features and patterns associated with the target structures.

We also assessed our hierarchical approach on the KiTS23 using the test set provided by the KiTS23 organizers. These results, calculated by the KiTS23 organizing team, are presented in Table 2.

After a comprehensive analysis of our initial submission to KITS23, we recognized certain limitations in our pipeline, as shown in Fig. 6. In light of these observations, we have conducted an additional experiment by deploying a 2D model for model K. Given the inherent characteristics of 2D convolutions, a 2D model can utilize the entire axial slice as input, as opposed to the small patch sizes used in 3D models. This reduces the likelihood of erroneously predicting extensive regions that are not truly representative of kidneys. It is imperative to note that our post-processing steps demonstrated efficacy during the validation phase, as evidenced by the validation results. We used a 2D U-Net architecture for model K. Table 3 showcases the outcomes derived from this model. As evident from the result, this methodology exhibits a marginal improvement over our earlier approach—i.e., 3D U-Net architecture for model K. The 2D model

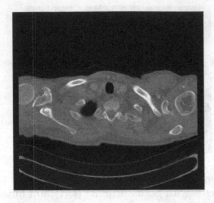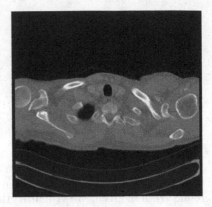

Fig. 6. Incorrect prediction made by 3D pipeline (left) and correct prediction from 2D pipeline (right). (Color figure online)

solved the issue in the 3D pipeline, as shown in Fig. 6. The measures in Table 3 were provided by the KiTS23 organizers based on our submissions.

Table 2. Our test set results with 3D models

	Kidney	Mass	Tumor	Mean
Dice	0.938	0.743	0.624	0.768
Surface Dice	0.892	0.602	0.490	0.661

Table 3. Comparison of Dice and Surface Dice Metrics for 2D vs. 3D

	Dice				Surface Dice			
	Kidney	Mass	Tumor	Overall	Kidney	Mass	Tumor	Overall
3D	0.938	0.743	0.624	0.768	0.892	0.602	0.490	0.661
2D	0.937	0.749	0.629	0.772	0.893	0.605	0.489	0.662

We trained our final models on two NVIDIA RTX A6000 GPUs. For model K, each epoch took, on average, 30 s to complete. For models M and T, the processing time for each epoch varied between 5 and 10 s. This substantial reduction in processing time was due to the large reduction in spacial size. We trained each model for 1000 epochs.

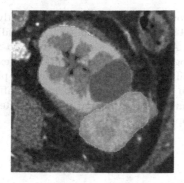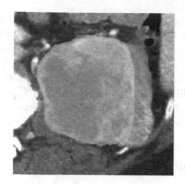

Fig. 7. Noise in the ground truth in case 102(left) and case 426(right). Kidney tissue is represented in green, tumor yellow, and cyst red. (Color figure online)

4   Discussion and Conclusion

We developed an efficient hierarchical deep-learning pipeline for the segmentation of kidneys, renal cysts, and renal tumors. We demonstrated the high performance of our approach both quantitatively and qualitatively.

Though we utilized three models in our pipeline, models M and T require little computational resources due to the isolation of the kidney data in model K, resulting in a substantial reduction in spatial dimension. Consequently, it requires little memory and computation to train these models. Additionally, cropping the kidney ROI allows for a larger and more diverse batch. In particular, we trained our 3d U-Net with a batch size of 2, and our 2d U-Net with a batch size of 99. We believe that the slight increase in computation that our pipeline introduces is justified by the performance gained in tumor and cyst segmentation. A notable limitation of this work pertains to the quality of the annotations used for model training, particularly regions of interest that have been missed or over-contoured during the manual annotation. Figure 7 shows two examples where small parts of the kidney are missed in the ground truth. Such inconsistencies within the training dataset pose challenges for training reliable and accurate segmentation models. These irregularities can impede the model's ability to learn and generalize effectively, potentially leading to sub-optimal performance during inference.

In conclusion, our hierarchical pipeline, which contains three models based on nnU-Net v2, demonstrates promising performance in segmenting kidneys, renal tumors, and renal cysts. We leverage the predicted kidney mask to refine the segmentation of tumors and cysts, effectively improving the accuracy of these classes. While our pipeline achieves high accuracy in kidney and mass segmentations, there is room for improvement in tumor and cyst predictions. Future work, including refining the manual segmentations or ensembling multiple predictions, can be performed to achieve better performance.

Acknowledgment. We would like to acknowledge the support of the Advanced Research Computing (ARC) cluster at the University of Calgary for providing us with access to their computational resources. Additionally, we are grateful to the AI Research School and Google's exploreCSR program for their contributions to expanding our knowledge in the field, which greatly influenced the development of this project. We would also like to thank the KiTS23 team for their support during and after the KiTS23 challenge. Lastly, we would like to thank the members of the Vision Research Lab at the University of Calgary for their support.

References

1. da Cruz, L.B., et al.: Kidney segmentation from computed tomography images using deep neural network. Comput. Biol. Med. **123**, 103906 (2020)
2. Hsiao, C.H., et al.: A deep learning-based precision and automatic kidney segmentation system using efficient feature pyramid networks in computed tomography images. Comput. Methods Programs Biomed. **221**, 106854 (2022)
3. Isensee, F., Jaeger, P.F., Kohl, S.A., Petersen, J., Maier-Hein, K.H.: nnU-Net: a self-configuring method for deep learning-based biomedical image segmentation. Nat. Methods **18**(2), 203–211 (2021)
4. Kittipongdaja, P., Siriborvornratanakul, T.: Automatic kidney segmentation using 2.5D ResUNet and 2.5D DenseUNet for malignant potential analysis in complex renal cyst based on CT images. EURASIP J. Image Video Process. **2022**(1), 5 (2022). https://doi.org/10.1186/s13640-022-00581-x
5. Ronneberger, O., Fischer, P., Brox, T.: U-Net: convolutional networks for biomedical image segmentation. In: Navab, N., Hornegger, J., Wells, W., Frangi, A. (eds.) MICCAI 2015. LNCS, vol. 9351, pp. 234–241. Springer, Cham (2015). https://doi.org/10.1007/978-3-319-24574-4_28
6. Siegel, R.L., Miller, K.D., Wagle, N.S., Jemal, A.: Cancer statistics, 2023. CA Cancer J. Clin. **73**(1), 17–48 (2023)
7. Thong, W., Kadoury, S., Piché, N., Pal, C.J.: Convolutional networks for kidney segmentation in contrast-enhanced CT scans. Comput. Methods Biomech. Biomed. Eng. Imaging Vis. **6**(3), 277–282 (2018)
8. Zhao, Z., Chen, H., Wang, L.: A coarse-to-fine framework for the 2021 kidney and kidney tumor segmentation challenge. In: Heller, N., Isensee, F., Trofimova, D., Tejpaul, R., Papanikolopoulos, N., Weight, C. (eds.) KiTS 2021. LNCS, vol. 13168, pp. 53–58. Springer, Cham (2022). https://doi.org/10.1007/978-3-030-98385-7_8

Cascade UNets for Kidney and Kidney Tumor Segmentation

Konstantinos Koukoutegos[1,2(✉)], Frederik Maes[3], and Hilde Bosmans[1,2]

[1] Department of Radiology, UZ Leuven, Herestraat 49, 3000 Leuven, Belgium
konstantinos.koukoutegos@uzleuven.be
[2] Department of Imaging and Pathology - Medical Physics and Quality Assessment,
KU Leuven, Herestraat 49, 3000 Leuven, Belgium
[3] Department of Electrical Engineering, KU Leuven, Kasteelpark Arenberg 10,
3001 Leuven, Belgium

Abstract. Kidney and kidney tumor delineation constitutes an essential and labor-intensive task performed by expert radiologists and clinicians in order to diagnose renal-related pathologies, optimize radiotherapy treatment and speedup surgical planning. In this work, we address the problem of kidney delineation using 3D UNets to automatically segment the kidneys and kidney tumors and cysts in Computed Tomography (CT) series. A 3-stage cascade approach of UNets was applied to first segment kidneys and masses at low resolution with a coarse network, followed by a fine segmentation step at the second stage. During this refinement step, the predicted labelmap from the first stage is used as shape prior knowledge in order to guide the training of the model. At last, two independent UNets segment separately masses and tumors using the prior knowledge of the refinement step. Post-processing operations consist of 3D connected component analysis in order to remove false positive voxels out of the final predicted labelmap. We train our networks and evaluate the effectiveness of this approach in the KiTS23 challenge and dataset.

Keywords: Kidney and kidney mass segmentation · Cascade UNets · Deep learning

1 Introduction

Renal cell carcinoma (RCC) was the ninth most common cancer in 2012 [4]. Partial or radical nephrectomy, deduction of the malignant tissue or active tumor surveillance are common ways to tackle RCC [3] and all of them require an accurate segmentation of the kidney and kidney tumor area. Manually delineating kidneys and kidney masses to obtain a precise segmentation is quite labor-intensive and time-consuming when performed by expert radiologists or clinicians. Additionally, manual segmentations tend to suffer from interobserver variability whilst deep learning-based approaches show more promising results

© The Author(s), under exclusive license to Springer Nature Switzerland AG 2024
N. Heller et al. (Eds.): KiTS 2023, LNCS 14540, pp. 107–113, 2024.
https://doi.org/10.1007/978-3-031-54806-2_15

in terms of consistency [7]. Therefore, automating the process of kidney and kidney tumor segmentation can provide significant benefits.

In this study, we develop a multi-stage segmentation approach inspired by Zhao et al. [8] to accurately segment kidneys and kidney masses, i.e. kidney tumors and kidney cysts, in CT images. During the multi-stage segmentation, firstly a coarse labelmap is generated and it is passed through to the downstream UNet as shape prior, to guide the fine segmentation step. Tumors and masses are segmented separately during the third step in order to obtain the final labelmap comprising kidneys, tumors and cysts.

2   Methods

Figure 1 outlines the structure of our method. Since the desired foreground voxels account for only a fraction of the whole CT series, we firstly segment the kidneys in a coarse manner in order to locate them. It is based on this coarse segmentation step that a region of interest (ROI) is generated from the original volume that is then passed to the downstream segmentation models where the initial segmentation map is refined and then separately segmented into tumors and cysts.

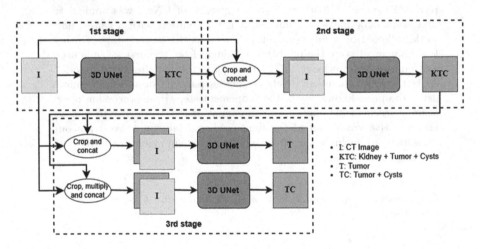

Fig. 1. Multi-stage cascaded approach for coarse to fine kidney and kidney masses segmentation

2.1   Training and Validation Data

During training, we made use only of the original dataset released as part of the KiTS23 challenge. This dataset is based on the one used for KiTS19 challenge [2], with additional cases included. It consists of 489 CT examinations, acquired during either late arterial or nephrogenic phase of contrast imaging. In order to select a good model architecture, an optimal set of hyperparameters and obtain a rough estimation of generalization error, we split the data into training, validation and test sets of 359, 65 and 65 cases respectively.

2.2 Preprocessing

Similar data preprocessing pipelines were used to train all four models of our method. HU values were normalized according to the winning submission of the KiTS19 challenge [1] for all stages. During the coarse segmentation step, images were resampled to 3 mm isotropic resolution. For the downstream fine segmentation task, images were resampled to 0.8 mm at off-plane resolution while the in-plane resolution was maintained according to the original pixel spacing. This choice seemed natural as the surface dice coefficient [6] is one of the metrics used to evaluate the predicted segmentations and we aimed for this network to correct the predicted contours as much as possible. During the final tasks of tumor and masses segmentation, images where resampled to 0.8 mm isotropic voxels. All networks were trained using randomly cropped patches from the CT volumes, and data augmentation techniques were used, including random flipping in all axes, random intensity shifting and addition of Gaussian noise.

2.3 Proposed Method

All four models of our method were identical and trained in a similar manner. The UNet architectures were based on [5] using additional residual blocks in each layer. At each encoding layer, we first downsampled the image using strided convolutions. The following part of the layer consists of 3 subsequent blocks of **Conv3D-BatchNorm-PRelu** followed by a residual skip connection. We did not downsample the image after the last encoding layer, before the bottleneck. Similarly at the decoding side, we first used transpose convolutions of size $3 \times 3 \times 3$ to upsample the image, followed by a single block of same structure as in the encoder layers. All convolution kernels are of size $3 \times 3 \times 3$.

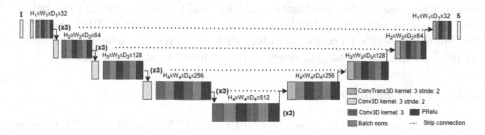

Fig. 2. UNet architecture

2.3.1 Coarse Segmentation

At the first stage of the proposed approach, the input to the UNet consists of a CT volume $I \in \mathbb{R}^{64 \times 64 \times 64}$ resampled to a 3 mm isotropic resolution. The UNet architecture is depicted in Fig. 2. The loss function used is the combination of Dice and Cross entropy, the batch size was set to 120 patches and the learning rate was kept at a constant value of 10^{-3} while no scheduling scheme was applied at this stage. A dropout rate of 0.4 was applied.

2.3.2 Fine Segmentation

At the second stage, the input to the UNet consists of a CT volume $I \in \mathbb{R}^{2 \times 96 \times 96 \times 96}$ resampled at 0.8 mm off-plane resolution while the in-plane pixel spacing was kept constant. The dual-channel input contains a cropped volume from the original CT around the kidney area, along with the generated segmentation map from the first stage which acts as shape prior knowledge, helping the network to quickly focus on the structures of interest. In this way, the main purpose of the second stage is to refine the boundaries of the coarse segmentation map and correct any false positives or false negatives that were predicted during the initial step. The UNet architecture is the same as in the first stage and depicted in Fig. 2. The loss function used is the combination of Dice and Cross entropy and the batch size was set to 100 patches. The learning rate was initially set to 10^{-3} and after 100 epochs it followed a cosine annealing scheme, decreasing to 10^{-5} during 50 epochs. The dropout rate applied to this stage was 0.3.

2.3.3 Tumor Segmentation

At the tumor segmentation stage, the input to the UNet consists of a CT volume $I \in \mathbb{R}^{2 \times 96 \times 96 \times 96}$ resampled at 0.8 mm isotropic resolution. At this point, the second channel of the input volume is the refined mask obtained during the second stage and the input volume is a cropped patch based on the refined mask. The UNet architecture is the same as in the first and second stages. The loss function used is the combination of Dice and Cross entropy and the batch size was set to 48 patches. The learning rate was initially set to 10^{-3} and after 200 epochs it followed a cosine annealing with warm restarts scheme, decreasing to 10^{-4} during 30 epochs. The dropout rate applied to this stage was 0.3.

2.3.4 Mass Segmentation

At the mass segmentation stage, the input to the UNet consists of a CT volume $I \in \mathbb{R}^{2 \times 96 \times 96 \times 96}$ resampled at 0.8 mm isotropic resolution. For this model, the dual-channel input contains the refined mask generated before, and a cropped patch from the original volume (based on this mask) with most of its background area masked out. The reasoning behind this masking is that upon visual inspection of the dataset we found that in some cases, parts of the cysts have similar texture with surrounding tissues. It is these similarities that we tried to eliminate by forcing the network to regard them as background. The UNet architecture is the same as in the first and second stages. The loss function used is the combination of Dice and Cross entropy and the batch size was set to 24 patches. The learning rate was initially set to 3×10^{-4} and after 200 epochs it followed a cosine annealing with warm restarts scheme, decreasing to 10^{-4} during 30 epochs. The dropout rate applied to this stage was 0.3.

2.4   Post Processing

As part of our post processing we performed 3D connected component analysis after the mask refinement step, in order to remove any false positives. Following the tumor and mass prediction, we removed false negatives by keeping only segmented parts that are located inside the refined mask. We then combined the tumors and masses predictions into a single segmentation map by keeping all remaining predicted voxels. Ultimately, we used trilinear interpolation to obtain a final segmentation map with the same resolution as the original CT image.

3   Results

Multiple models and multiple set-ups and post-processing pipelines were evaluated in order to pick the best performing model for the challenge. Table 1 summarizes quantitative results on $test\_set\_1$, which originates directly from the train, eval, test split performed on the dataset released publicly by the challenge organizers. Model crf_model utilizes a Conditional Random Field (CRF) at the end of the post-processing mentioned in Sect. 2.4 in order to refine the segmentation output. Models $overlap\_0.50$ and $overlap\_0.75$ utilize an ensemble prediction scheme based on overlapping windows during inference time with 50% and 75% overlap respectively. It can be observed that the crf_model demonstrates similar performance when used to refine the KTC mask while for TC and T separately the Dice and Surface Dice scores dropped.

Table 1. Hierarchical class evaluation for distinct trained models.

Model	Dice KTC	Dice TC	Dice T	SDice KTC	SDice TC	SDice T
crf_model	0.967	0.780	0.726	0.930	0.622	0.571
$overlap\_0.50$	0.968	**0.790**	**0.736**	0.931	**0.633**	**0.583**
$overlap\_0.75$	**0.971**	0.789	**0.736**	**0.934**	**0.633**	0.582

Model $overlap\_0.75$ was used to segment the hidden test set's cases and the results can be found in Table 2. Our overall method ranked 17th on the official KiTS23 leaderboard.

Table 2. Results of our proposed method on the official KiTS23 leaderboard.

Model	Dice KTC	Dice TC	Dice T	SDice KTC	SDice TC	SDice T
*overlap*_0.75	0.939	0.728	0.643	0.885	0.561	0.482

Figure 3 demonstrates two cases of *test_set*_1, for which the ground truth segmentation map is known. It can be observed that the overall mask for kidneys and masses (KTC) shows acceptable agreement when compared with the ground truth segmentation map. On the contrary, masses and tumors masks (TC and T) can be over or under-segmented in some cases, especially when the surrounding tissue HU values are similar to the kidney ones.

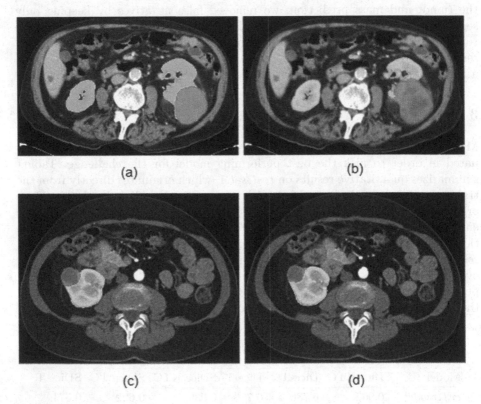

Fig. 3. Ground truth (a), (c) and predicted (b), (d) segmentation maps.

Every stage of the proposed method was trained independently from each other and all experiments utilized 3 NVIDIA GeForce GTX 1080 Ti with 11 GB of memory. All models were trained for 1500 epochs without early stopping and each model took 4 days to train. Inference was performed in a single NVIDIA GeForce RTX 2080 Ti with 11 GB of memory and each case took approximately 45 s.

4   Discussion and Conclusion

In this paper, we proposed a cascaded approach to address the segmentation of kidneys and kidney masses in CT images acquired during either late arterial or nephrogenic phases of contrast imaging. Our method is based on first coarsely segmenting the kidneys and kidney masses together in a low-resolution setup, then refine the segmentation mask during the second stage to obtain a high resolution segmentation map. Lastly, two separate networks were used to segment tumors and masses in the third stage. The overall approach was implemented using the official KiTS23 dataset and upon validation with the challenge's hidden test set, our method ranked 17th on the KiTS23 leaderboard.

References

1. Heller, N., et al.: The state of the art in kidney and kidney tumor segmentation in contrast-enhanced CT imaging: results of the KiTS19 challenge. Med. Image Anal. **67**, 101821 (2021)
2. Heller, N., et al.: The KiTS19 challenge data: 300 kidney tumor cases with clinical context, CT semantic segmentations, and surgical outcomes. arXiv preprint arXiv:1904.00445 (2019)
3. Hsieh, J.J., et al.: Renal cell carcinoma. Nat. Rev. Dis. Primers **3** (2017). https://doi.org/10.1038/nrdp.2017.9
4. Jonasch, E., Gao, J., Rathmell, W.K.: Renal cell carcinoma. BMJ **349** (2014). https://doi.org/10.1136/bmj.g4797
5. Kerfoot, E., Clough, J., Oksuz, I., Lee, J., King, A.P., Schnabel, J.A.: Left-ventricle quantification using residual U-Net. In: Pop, M., et al. (eds.) STACOM 2018. LNCS, vol. 11395, pp. 371–380. Springer, Cham (2019). https://doi.org/10.1007/978-3-030-12029-0_40
6. Nikolov, S., et al.: Deep learning to achieve clinically applicable segmentation of head and neck anatomy for radiotherapy (2018). http://arxiv.org/abs/1809.04430
7. Webb, J.M., et al.: Comparing deep learning-based automatic segmentation of breast masses to expert interobserver variability in ultrasound imaging. Comput. Biol. Med. **139**, 104966 (2021). https://doi.org/10.1016/j.compbiomed.2021.104966
8. Zhao, Z., Chen, H., Wang, L.: A coarse-to-fine framework for the 2021 kidney and kidney tumor segmentation challenge. In: Heller, N., Isensee, F., Trofimova, D., Tejpaul, R., Papanikolopoulos, N., Weight, C. (eds.) KiTS 2021. LNCS, vol. 13168, pp. 53–58. Springer International Publishing, Cham (2022). https://doi.org/10.1007/978-3-030-98385-7_8

Cascaded nnU-Net for Kidney and Kidney Tumor Segmentation

Yaqi Wang, Yu Dai$^{(\boxtimes)}$, Jianxun Zhang, and Jingjing Yin

College of Artificial Intelligence, Nankai University, Tianjin, China
daiyu@nankai.edu.cn

Abstract. Kidney cancer has been recognized as one of the top ten prevalent neoplastic conditions, ranking as the third most frequent malignant tumor within the genitourinary system. Its high mortality rates pose a significant risk to human health. Accurate and automated segmentation of the kidneys, kidney tumors, and kidney cysts in CT scans is of paramount importance, as it provides medical professionals with valuable assistance in their diagnostic and therapeutic efforts. In KiTS23, this work presents a novel two-stage cascaded framework based on the nnU-Net architecture. Between the two stages of the cascaded network, a cropping process is implemented. This process involves extracting a region of interest (ROI) that encompasses the kidneys from the initial segmentation. The extracted ROI is subsequently utilized as input for the second stage, facilitating more focused and refined segmentation of kidney tumors and cysts. Furthermore, to tackle the inherent challenge of class imbalance, Focal Loss is employed as a mitigation strategy. The network achieved average Sørensen-Dice scores of 0.933, 0.709, and 0.645 for the classes kidney, masses and tumor respectively. Similarly, the average surface Dice scores for these classes were 0.866, 0.545, and 0.490. This led to the 18th position in the KiTS23 challenge.

Keywords: nnU-Net · cascaded network · imbalanced segmentation

1 Introduction

According to the Global Cancer Statistics 2022, the prevalence of kidney tumors surpasses 420,000 individuals, resulting in approximately 180,000 deaths [2]. Accurate delineation of tumor boundaries provides medical professionals with essential information regarding size, shape, volume, and spatial characteristics. This information aids in diagnosis, treatment planning, and monitoring tumor progression. It significantly influences clinical decision-making, allowing tailored treatment strategies for individual patients.

Manual segmentation of kidney tumors is a time-consuming and laborious task, compounded by potential inter-observer variability arising from subjective perceptions among different medical professionals. Therefore, automated and precise kidney tumor segmentation holds immense importance in the field of medical image analysis.

© The Author(s), under exclusive license to Springer Nature Switzerland AG 2024
N. Heller et al. (Eds.): KiTS 2023, LNCS 14540, pp. 114–119, 2024.
https://doi.org/10.1007/978-3-031-54806-2_16

Segmenting small tumors within a large background poses challenges. This paper primarily focuses on mitigating the issue of class imbalance in medical image segmentation. A two-stage cascaded framework based on the nnU-Net [4] architecture is proposed. In the initial stage, a 3D full-resolution U-Net model accurately delineates the kidney region and extracts the region of interest (ROI). Subsequently, another 3D full-resolution U-Net model is employed in the following stage to further segment kidney tumors and cysts within the extracted ROI. To address the challenge of class imbalance, Focal Loss is utilized as a mitigation strategy.

2   Methods

The proposed method focuses on addressing the issue of class imbalance, and it primarily involves a two-stage segmentation approach using two 3D full-resolution U-Net models based on nnU-Net. The kidneys, tumors and cysts are segmented in separate stages, with an intermediate cropping step to extract a Region of Interest (ROI) containing the kidneys. Finally, an ensemble method is employed to generate the segmentation results. See Fig. 1.

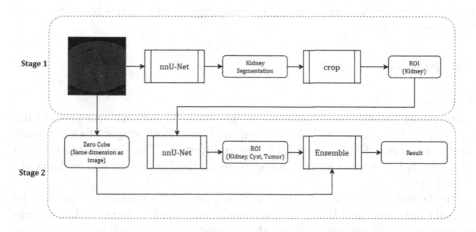

Fig. 1. Network Architecture

The key of the method can be summarized as follows:

1. A cropping technique is applied to mitigate the volume disparity between the kidneys and tumors/cysts, thereby reducing the impact of smaller tumor and cyst volumes on the segmentation of the kidneys and background.
2. The Focal loss function [5] is utilized, which incorporates a weighting mechanism based on the difficulty of segmentation. This weighting strategy helps address the challenges posed by class imbalance.

2.1 Training and Validation Data

Our submission made use of the official KiTS23 [1] training set alone.

2.2 Preprocessing

In the data preprocessing phase, a pipeline configuration method based on nnU-Net is employed, incorporating crop, re-sampling, and HU normalization procedures. Initially, the training set was cropped to retain solely the non-zero regions.

Given the presence of anisotropy in the KITS23 data, resampling was performed to achieve a target spacing of [0.98, 0.78125, 0.78125]. Subsequently, statistical analysis was conducted on the foreground information of the entire CT image training set. Specifically, the Hounsfield Unit (HU) values of each image data are clipped to the range of the 0.5 to 99.5 percentiles, computed from all foreground voxels. Additionally, mean and standard deviation (s.d.) are calculated based on the global foreground information for subsequent Z-score standardization.

Moreover, a series of data augmentation operations are integrated into the training process. These operations, including rotations, scaling, Gaussian noise, Gaussian blur, brightness adjustments, low-resolution simulation, gamma adjustments, and mirroring, were probabilistically applied to enhance the diversity of the training samples.

2.3 Proposed Method

Network Architecture

The proposed method employs a two-stage cascaded network structure, utilizing a 3D full-resolution U-Net based on nnU-Net. Both stages of the network employ the same 3D full-resolution U-Net architecture. This design aims to avoid information loss caused by resampling between the two stages due to disparate resolutions.

In the first stage, following the aforementioned preprocessing operations, the first 3D full-resolution U-Net architecture is employed to perform kidney segmentation on CT images. Utilizing the obtained segmentation results, extreme coordinates in six directions are determined, defining the minimum cubic region that encompasses the kidneys. This region is further expanded by a factor of 1.5 [6] to acquire a Region of Interest (ROI) that fully encapsulates the kidneys. Subsequently, the ROI is passed as an input into the second stage.

In the second stage, the same 3D full-resolution U-Net architecture based on nnU-Net is used to further segment the kidneys, tumors, and cysts. Finally, an ensembling process is applied to combine the segmentation results. By utilizing the recorded positional information, the results are reconstructed in the original image, yielding the final segmentation outcome.

Loss Function

In the first stage, the Dice Loss and Cross-Entropy Loss functions are employed.

In the second stage, the Focal Loss and Cross-Entropy Loss functions are employed.

Optimization Strategy

During the network initialization phase, to accelerate the convergence speed, we utilize pre-training weights from "Task135_KiTS2021" on nnU-Net. This choice is based on the fact that the initial 300 cases of the KiTS23 dataset are identical to the KiTS21 dataset, and our training set already includes all of these cases. By incorporating the pre-training weights, the objective was to improve the network's convergence rate and facilitate efficient learning.

During the training process, the models are trained using the stochastic gradient descent (SGD) optimizer for a total of 450 epochs. Each epoch comprises 250 batches, with a batch size of 2. The initial learning rate is set to 0.01, and it is decayed according to the function $0.01 \times (1 - \frac{epoch}{max\ epochs})^{0.9}$. The patch size for both stages is set to [128, 128, 128]. Considering the scarcity of expert-annotated medical image data and the extensive training time associated with nnU-Net's five-fold cross-validation, we aim to maximize the utilization of available cases for network training while minimizing the time consumption. As a result, the training data is not partitioned into separate training and validation sets. Instead, the entire training dataset is exclusively utilized for model training, reserving the test set solely for the purpose of evaluating the network's segmentation performance. Furthermore, all other hyperparameters follow the default settings of nnU-Net.

3 Results

The proposed models are implemented using nnU-Net framework with Python 3.7 and PyTorch framework on NVIDIA RTX3090 GPU with 25.4 GB Memory.

A total of 489 cases were included in KiTS23 dataset, with 450 cases used for training and 39 cases reserved for testing purposes. The segmentation results for each class is evaluated using the Sørensen-Dice and Surface Dice coefficient. On the test set, we use the official test program [3] of KiTS23 and the network achieved Sørensen-Dice scores are 0.921, 0.727, and 0.708 for the kidney, masses and tumor respectively. Additionally, the mean Surface Dice scores for the same classes were 0.854, 0.573, and 0.544, respectively. Detailed test results are presented in Table 1.

In the evaluation conducted on the additional 100 test sets provided by KiTS23 official dataset, our network achieved an average Sørensen-Dice of 0.763, an average Surface Dice of 0.634, and a tumor Dice of 0.645. Detailed official results are presented in Table 2.

Some of the results predicted by the model are illustrated in Fig. 2.

Table 1. Experiment results on the test set (splitted by ourselves)

	Kidney	Masses	Tumor	Average
Dice	0.921	0.727	0.708	0.785
Surface Dice	0.854	0.573	0.544	0.657

Table 2. Official results from the leaderboard

	Kidney	Masses	Tumor	Average
Dice	0.933	0.709	0.645	0.763
Surface Dice	0.866	0.545	0.490	0.634

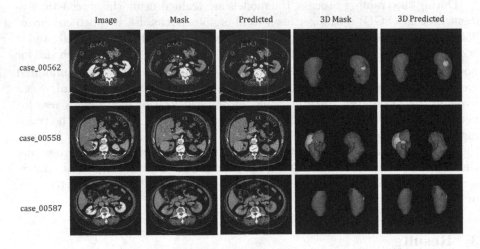

Fig. 2. Segmentation results from our Network

However, there are some cases where the segmentation results are not satisfactory due to certain reasons. For instance, in case 582, an extremely rare morphology of the kidney was observed, known as horseshoe kidney, where the two kidneys are fused together and form a U-shaped structure. See Fig. 3.

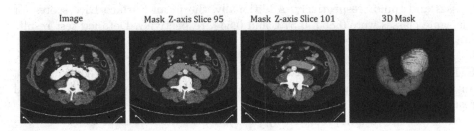

Fig. 3. Unusual anomaly observed in case_00582

For example, in case 586, the symptom pattern of the patient is significantly different from the majority. The volume of the cysts is much larger than that of the tumors, and the majority of the cysts are located outside the kidneys. Unfortunately, our segmentation results erroneously classified some of the cysts as tumors. See Fig. 4.

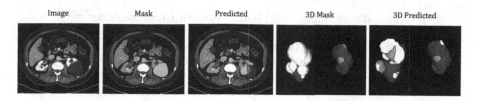

Fig. 4. Unusual example and our segmentation results

4   Discussion and Conclusion

In the work, a novel two-stage cascaded framework based on the nnU-Net architecture was proposed to address the challenge of accurate and automated segmentation of kidneys, kidney tumors, and kidney cysts in CT scans. The key contributions of the proposed method were the implementation of a cropping technique to mitigate volume disparity and the utilization of Focal Loss to address class imbalance. The average Dice and Sørensen-Dice coefficients for the segmentation results are 0.785 and 0.657, respectively.

As newcomers in this field and facing time constraints, there is ample room for improvement in our results. We would like to express our gratitude to the organizers for providing us with this valuable opportunity. Although our results may not be flawless, our participation has allowed us to gain valuable insights and knowledge in the process.

References

1. The 2023 kidney and kidney tumor segmentation challenge. https://kits-challenge.org/kits23/
2. Chhikara, B.S., Parang, K.: Global Cancer Statistics 2022: the trends projection analysis. Chem. Biol. Lett. **10**(1), 451–451 (2023)
3. Heller, N., et al.: The KiTS21 challenge: automatic segmentation of kidneys, renal tumors, and renal cysts in corticomedullary-phase CT (2023)
4. Isensee, F., Jaeger, P.F., Kohl, S.A., Petersen, J., Maier-Hein, K.H.: nnU-Net: a self-configuring method for deep learning-based biomedical image segmentation. Nat. Methods **18**(2), 203–211 (2021)
5. Lin, T.Y., Goyal, P., Girshick, R., He, K., Dollár, P.: Focal loss for dense object detection. In: Proceedings of the IEEE International Conference on Computer Vision, pp. 2980–2988 (2017)
6. Zhao, Z., Chen, H., Wang, L.: A coarse-to-fine framework for the 2021 kidney and kidney tumor segmentation challenge. In: Heller, N., Isensee, F., Trofimova, D., Tejpaul, R., Papanikolopoulos, N., Weight, C. (eds.) KiTS 2021. LNCS, vol. 13168, pp. 53–58. Springer, Cham (2022). https://doi.org/10.1007/978-3-030-98385-7_8

A Deep Learning Approach
for the Segmentation of Kidney, Tumor
and Cyst in Computed Tomography Scans

Kartik Narayan Sahoo[1] and Kumaradevan Punithakumar[2]([✉])

[1] Indian Institute of Technology Kharagpur, Kharagpur, India
kartik.sahoo@kgpian.iitkgp.ac.in
[2] Radiology and Diagnostic Imaging, University of Alberta, Edmonton, Canada
punithak@ualberta.ca

Abstract. Kidney cancer, also known as renal cell carcinoma, is a malignant tumor that originates in the kidneys. It is one of the most common types of cancer affecting the urinary system. Kidney tumors can vary in size, location, and aggressiveness, making early detection and accurate diagnosis crucial for effective treatment planning. The proposed method makes use of nnU-Net which is a self-adapting semantic segmentation method, to segment the kidney, tumor and cyst. The proposed neural network model was trained using the datasets provided by the 2023 Kidney and Kidney Tumor Segmentation Challenge hosted by MICCAI 2023 conference. The proposed methodology leveraged the power of deep learning to yield high segmentation accuracy.

Keywords: computed tomography scans · deep convolutional neural network · kidney · tumor · cyst · image segmentation

1 Introduction

Kidney cancer, also known as renal cell carcinoma (RCC), is one of the most prevalent malignancies worldwide, with more than 330,000 new cases being diagnosed annually [6]. The number of cases for kidney tumors have been increasing since the past few decades [2]. It is characterized by the uncontrolled growth of abnormal cells within the kidney. Accurate and precise segmentation of kidney tumors and cysts plays a crucial role in the diagnosis, treatment planning, and monitoring of kidney cancer [7]. In recent years, deep learning techniques, such as the nnU-Net framework [3], have shown remarkable potential in medical image segmentation tasks. nnU-Net is a state-of-the-art deep convolutional neural network architecture that has been successfully applied to various medical imaging tasks. The framework leverages a cascaded U-Net architecture [5], which consists of multiple nested U-Net subnetworks. The network is trained using a combination of dice and cross-entropy loss functions, with extensive data augmentation techniques to enhance robustness and generalization. nnU-Net has demonstrated

ⓒ The Author(s), under exclusive license to Springer Nature Switzerland AG 2024
N. Heller et al. (Eds.): KiTS 2023, LNCS 14540, pp. 120–125, 2024.
https://doi.org/10.1007/978-3-031-54806-2_17

remarkable success in various medical imaging applications, including segmentation of organs, tumors, lesions, and abnormalities. Its flexibility, adaptability, and superior performance make it a valuable tool for precise and accurate medical image segmentation tasks. This paper presents an approach for kidney, tumor, and cyst segmentation through deep convolutional neural networks (CNNs), using the nnU-Net architecture. The proposed methodology aims to leverage the power of deep learning to achieve accurate and robust segmentation of the kidney, tumor and cyst structures in medical images, particularly in computed tomography (CT) scans. The proposed neural network utilized for this challenge was trained on a dataset consisting of 489 cases of patients who underwent cryoablation, partial nephrectomy, or radical nephrectomy for suspected renal malignancy. These cases were collected from the years 2010 to 2022 at a M Health Fairview medical center. The CT scan dataset was provided by the 2023 Kidney and Kidney Tumor Segmentation Challenge organizers.

2   Methods

The complete workflow, encompassing both the training and inference stages, is visually illustrated in Fig. 1. Our segmentation approach for kidney, tumor, and cyst regions employed the nnU-Net architecture, without any modifications or adaptations.

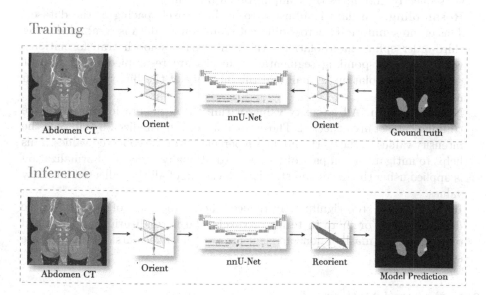

Fig. 1. The proposed neural network based solution to segment the kidney, tumor and cyst from abdomen CT scans.

2.1   Training and Validation Data

Our submission made use of the official KiTS23 training set alone. The dataset is composed of 599 cases with 489 allocated to the training set and 110 in the test set. Only the training set images and ground truths were available, whereas the test set images and ground truths were not revealed to the challenge participants. The challenge training set data (489 cases) was split to 391 training and 98 validation cases for the model. The CT scans are saved as 3D volumes and the dimension range from $(512 \times 512 \times 29)$ to $(512 \times 512 \times 1059)$. The annotated ground truths contain labels comprised of the kidney, tumor and cyst.

2.2   Preprocessing

The dataset's header information containing the position and orientation details of the 3D volume, was removed before the preprocessing step, as we found it to give improved model performance. The preprocessing method involves using the pipeline built with in the nnU-Net architecture. The steps carried out are as follows:

1. **Cropping.** Data undergoes cropping to regions of non-zero values. This cropping process is particularly beneficial as it reduces the size of the data and subsequently minimizes the computational burden.
2. **Resampling.** All data is adjusted to median voxel spacing of the dataset. This ensures uniformity across different scans. Image data is resampled using third-order spline interpolation, which allows for smooth transformations, while the corresponding segmentation masks are resampled using nearest neighbor interpolation to maintain the integrity of the binary segmentation information.
3. **Normalization.** All intensity values within the segmentation masks of the training dataset are collected. The entire dataset is normalized by clipping the intensity values to the 0.5th and 99.5th percentiles of the collected values. This helps to mitigate the impact of outliers. Additionally, a z-score normalization is applied using the mean and standard deviation of all the collected intensity values.
   If the cropping step significantly reduces the average size of patients in the dataset by 1/4 or more in terms of voxels, the normalization is performed only within the mask of nonzero elements and all values outside the mask are set to 0.

2.3   Proposed Method

The model is trained from scratch and evaluated using 5-fold cross validation on the training set. The network uses a combination of dice and cross-entropy loss as the loss function [3].

$$\mathcal{L}_{total} = \mathcal{L}_{dice} + \mathcal{L}_{CE}$$

In our optimization strategy, we employ the Adam optimizer with an initial learning rate of 3×10^{-4} for all experiments. To ensure efficient learning, we monitor the exponential moving average of the training loss. If there is no improvement in this loss for 30 epochs, we adjust the learning rate by reducing it by a factor of 5. If the exponential moving average of the validation loss does not improve by more than 5×10^{-3} within the last 60 epochs and the learning rate drops below 10^{-6}, the training process is stopped.

To prevent overfitting, the nnU-Net performs a variety of data augmentation techniques during training, which includes random rotations, random scaling, random elastic deformations, gamma correction augmentation and mirroring.

To increase the stability of the network, patch sampling is done, where a third of the samples in a batch have atleast one randomly chosen foreground class.

The neural network is trained for 1000 epochs, where an epoch is the iteration over 250 training batches. The training took around 3 days ($\sim$70 h) on the dataset using NVIDIA Tesla A100 (40 GB memory) GPU.

3   Results

The proposed method was quantitatively evaluated over validation CT dataset from over 98 patients. The validation set was derived from the original training set, and the ground truth annotations were available. Evaluation criteria in this research study were based on a method called "Hierarchical Evaluation Classes" (HECs) employed by the organizers. HECs involve combining classes that are subsets of another class to compute metrics for the superset. The HECs used in this study were as follows:

1. **Kidney and Masses**, which included Kidney, Tumor, and Cyst
2. **Kidney Mass**, comprising Tumor and Cyst
3. **Tumor**, focusing solely on Tumor segmentation

Evaluation metrics being used are the Sørensen-Dice and Surface Dice [4]. The class-wise dice scores are shown below:

Table 1 presents the average Sørensen-Dice and Surface Dice values obtained on the validation set of CT scans. The algorithm achieved Sørensen-Dice values of 97.48%, 86.82%, and 84.86% for the kidney and masses, kidney mass, and tumor HECs, respectively. The Surface-Dice values were similar with 96.70%, 77.97% and 73.98% respectively.

Table 2 presents the average Sørensen-Dice and Surface Dice values obtained on the test set of CT scans. The algorithm achieved Sørensen-Dice values of 91.8%, 68.5%, and 60.0% for the kidney and masses, kidney mass, and tumor HECs, respectively. The Surface-Dice values were 84.6%, 53.3% and 45.4% respectively.

The dice and surface-dice score overall were 0.734 and 0.611 respectively.

Table 1. The performance of the proposed algorithm on the validation CT datasets in terms of the Sørensen-Dice and Surface Dice metric. The table reports mean evaluation metrics for each of the HECs on the validation set as defined by the organizers.

Evaluation Metric	Kidney and Masses HEC	Kidney Mass HEC	TumorHEC
Sørensen-Dice (%)	97.485	81.933	78.272
Surface Dice (%)	96.709	77.979	73.989

Table 2. The performance of the proposed algorithm on the test CT datasets in terms of the Sørensen-Dice and Surface Dice metric. The table reports mean evaluation metrics for each of the HECs on the validation set as defined by the organizers.

Evaluation Metric	Kidney and Masses HEC	Kidney MassHEC	TumorHEC
Sørensen-Dice (%)	91.8	68.5	60.0
Surface Dice (%)	84.6	53.3	45.4

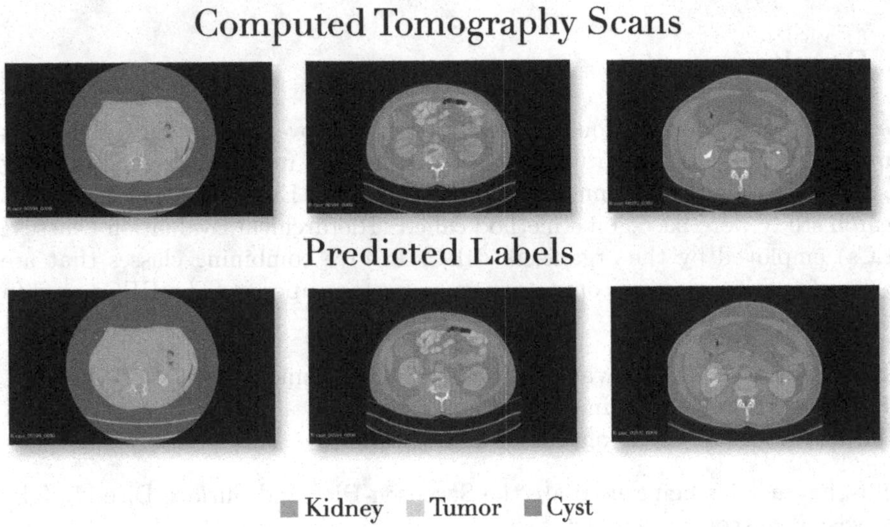

Fig. 2. Segmentation results on some test set abdomen CT scan images.

4   Conclusion

In this research study, we employed an nnU-Net approach based on deep convolutional neural networks to automatically segment the kidney, tumor, and cyst regions in CT scans. The proposed methodology was evaluated on a validation dataset comprising scans from 98 patients. To assess the performance, we converted the ground truth and predicted images into the three hierarchical evaluation classes (HECs) and employed Deepmind's Surface Distance library for evaluation metrics. The results demonstrated a strong agreement between the automated predictions and manual delineations, as indicated by Sørensen-Dice coefficient and Surface Dice values. Moving forward, our future work will be directed towards further improving the model's performance, specifically focusing on enhancing the dice score for cyst segmentation. This could be achieved by implementing a nested nnU-Net architecture, utilizing dedicated sub-networks for segmenting each individual component [1].

Acknowledgment. The authors would like to express their gratitude to the challenge organizers for providing the train dataset. It is hereby stated by the authors of this research paper that the implemented segmentation method, utilized for participation in the Kits23 challenge, did not make use of any pre-trained models or supplementary datasets beyond those provided by the organizers. The study received support from Mitacs as part of the Globalink Research Internship program in 2023. Furthermore, the authors acknowledge the computing resources made available by Digital Research Alliance of Canada (https://alliancecan.ca) and WestGrid, which facilitated the execution of the research.

References

1. Heller, N., et al.: The KiTS21 challenge: automatic segmentation of kidneys, renal tumors, and renal cysts in corticomedullary-phase CT (2023)
2. Hollingsworth, J.M., Miller, D.C., Daignault, S., Hollenbeck, B.K.: Rising incidence of small renal masses: a need to reassess treatment effect. JNCI: J. Natl. Cancer Inst. **98**(18), 1331–1334 (2006). https://doi.org/10.1093/jnci/djj362
3. Isensee, F., et al.: nnU-Net: self-adapting framework for U-Net-based medical image segmentation (2018)
4. Nikolov, S., et al.: Deep learning to achieve clinically applicable segmentation of head and neck anatomy for radiotherapy (2021)
5. Ronneberger, O., Fischer, P., Brox, T.: U-Net: convolutional networks for biomedical image segmentation. In: Navab, N., Hornegger, J., Wells, W., Frangi, A. (eds.) MICCAI 2015. LNCS, vol. 9351, pp. 234–241. Springer, Cham (2015). https://doi.org/10.1007/978-3-319-24574-4_28
6. Scelo, G., Larose, T.L.: Epidemiology and risk factors for kidney cancer. J. Clin. Oncol. **36**(36), 3574–3581 (2018). https://doi.org/10.1200/jco.2018.79.1905
7. Yang, G., et al.: Automatic kidney segmentation in CT images based on multi-atlas image registration. In: 2014 36th Annual International Conference of the IEEE Engineering in Medicine and Biology Society, pp. 5538–5541 (2014). https://doi.org/10.1109/EMBC.2014.6944881

Recursive Learning Reinforced by Redefining the Train and Validation Volumes of an Encoder-Decoder Segmentation Model

Antonio Vispi[✉]

Turin Polytechnic, Turin, Italy
antoniovispi1@gmail.com

Abstract. The following short paper is an attempt to automate the CT image segmentation process, through a Deep Learning-based approach, with the aim of segmenting the kidney and its possible pathological masses as accurately as possible. It was decided to use a segmentation model of the Encoder-Decoder type, it was decided to use an EfficientNet-B5 as encoder, and an Unet as decoder, suitably set up and modified. It was decided to perform several cascade trainings of the model, which will be called rounds, at the beginning of each of which a refined redefinition of the training and validation images was set up, allowing the model to deal with a large amount of data, increasing its generalisation capacity. Finally, following a careful search for the best training configurations of the models and the various training rounds, good results were obtained on the test set, with a segmentation accuracy of the Kidney + Tumor + Cyst, Tumor + Cyst, Tumor of 97.71%, 81.39% and 73.81% respectively.

Keywords: Segmentation · TensorFlow · Keras

1 Introduction

Kidney cancer is the most common malignancy of the kidney, one of the most common cancers today and with highest mortality rate. Generally, the diagnosis is confirmed based on computed tomography (CT) or magnetic resonance imaging (MRI), occasionally with biopsy. Detecting the possible presence of lesions is a key step in assessing a targeted treatment with greater likelihood of success for the patient. Segmentation methods are widely used in this field as they make it possible to identify the organ of interest and, eventually, malignant, or abnormal bodies. Among the different ways one could consider manually segmenting the images, as shown below (Fig. 1):

Note that each colour in the segmentation, whether manual or automatic, corresponds to a precise class. Shown below is the legend establishing the class-colour association, which will remain the same throughout the whole paper (Table 1).

Manual segmentation, while being a high-precision method, has significant disadvantages both in terms of timing and operator dependence. There is therefore a clear need for an automatic segmentation method that overcome these limitations.

© The Author(s), under exclusive license to Springer Nature Switzerland AG 2024
N. Heller et al. (Eds.): KiTS 2023, LNCS 14540, pp. 126–138, 2024.
https://doi.org/10.1007/978-3-031-54806-2_18

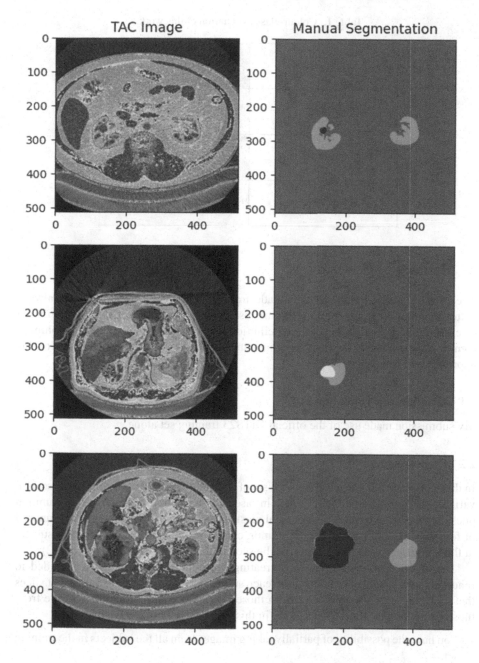

Fig. 1. Examples of abdominal CT images and corresponding manual segmentation.

Table 1. Colour-class association of this work.

Background	
Kidney	
Tumor	
Cyst	

2 Methods

The method adopted consists of cascade trainings, where the next model is always better than the previous one in terms of loss on the validation set. At the beginning of each training session, there is a redefinition of the training and validation volumes, significantly increasing the number of new images the model sees. More details of this approach will be given below.

2.1 Training and Validation Data

My submission made use of the official KiTS23 training set alone.

2.2 Preprocessing

In this work each volume represents the set of images of a subject and consists of several variable slices with a size of 512×512; in case of dimensions different from the common ones, a resize is performed. The manual mask associated with each image is composed of four classes: background, kidney, tumor, cyst. The number of slices for each subject in the CT images varies widely, as can be seen in the following histogram (Fig. 2):

For this reason, in the phase of creating the training volume, it was decided to randomly take a number of slices from each subject proportional to the number of slices that each subject has: taking a lot from those who have a lot of slices and a little from those who have few slices. Proceeding in this way has a twofold advantage:

- You have the possibility of partially taking images from all the subjects in the training set.
- Each time you redefine the train and validation volume, it will be very different from time to time.

This happens because the train and validation slices are taken randomly and uniquely; therefore, in the train volume the images are all different, and furthermore if I redefine the train and validation volume again, it will be very different from the previous one.

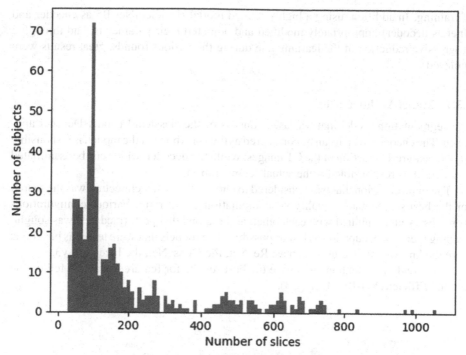

Fig. 2. Histogram of numerosity of slices per subject in the whole dataset.

This mechanism lends itself very well to cascading training, redefining the train volumes each time with each training round: in this way the model, after the various training rounds, gets to see a very large number of images from the dataset, getting better and better.To reduce the over-fitting of the model that leads to a loss in the ability to generalize and adapt to new situations, data augmentation of the training set was carried out. New images were obtained by applying the following operations to the original training images:

- Random rotations in a maximum range of 23°.
- Random zoom of the images in a range of 28%.
- Randomly changed the width and height dimensions of the images in a range of 21%.
- Randomized horizontal flipping of images.

Finally, it should be noted that no normalisations or subject selections were performed, neither in the train nor in the test phase, not to compromise generalisation capacity.

2.3 Proposed Method

Through a complete redefinition of the train and validation volume at each successive round (each round 8000 new images for the training volume and 1200 new images for the validation volume), the model gets to see a large amount of data, promoting robustness of training. In addition, using a highly refined model (EfficientNet-B5 as encoder and Unet as decoder) appropriately modified and adjusted their parameters, and through a progressive reduction of the learning rate during the various rounds, great results were achieved.

2.3.1 Model Architecture

The segmentation model that was used consists of the classical Encoder-Decoder app-roach. The encoder, which can be considered as the Backbone of the model, is responsible for the Feature-Extraction of the CT images; while the decoder, which can be assimilated to the head, is responsible for the actual segmentation.

The main criterion that was considered to choose the best architecture was the search for the best segmentation quality vs. computational cost ratio. Various configurations were observed, combined with each other, and the one that performed best was sought, keeping memory occupation as low as possible. The models that were tested as backbone were the known architectures such as: ResNet, the DenseNet, the Inception-v3.

In the end, the backbone that gave the best results for feature extraction fell on the famous **EfficientNet-B5** [3] (Fig. 3).

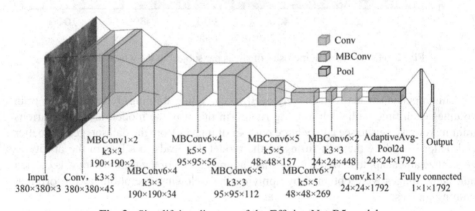

Fig. 3. Simplifying diagram of the EfficientNet-B5 model.

The MBConv operation is nothing more than a combination of operations that sim-plify the visualisation of the architecture. Note that the EfficientNet-B5 was trained from scratch.

Regarding the decoder (also trained from scratch), several architectures were tried, including Linknet, FPN, PSPNet. Finally, for the same reasons just explained, the best results were obtained from the **Unet** architecture, therefore it was chosen [1] (Fig. 4).

Furthermore, the decoder was modified, increasing the number of filters, to have a more detailed fit to the data distribution, increasing the final performance of the model. In particular, the number of filters used for the Unet that gave the best results was 512, 256, 128, 64, 32.

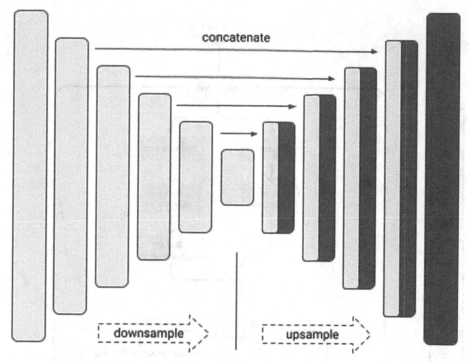

Fig. 4. Simplifying diagram of the Unet model.

2.3.2 Training Schedule

At this point, the details of the organisation of the various training sessions are shown. Let's say we run six successive training rounds of the model: the train and validation volume will be defined six different times. Clearly, for each training round, the model that forms the initialisation of the next round it is which one that performs better on the previous round, in terms of loss calculated on the validation set. In the end, the model that will be taken as the best of all will be the one with the lowest loss calculated on the validation set.

The pattern of cascading trainings, one after the other, after a redefinition of the train and validation volume at each round is summarised in the following flow chart (Fig. 5):

It should be clear that the overall training is the result of several rounds performed one after the other. Regarding individual rounds, it was decided to use the NADAM optimiser, or Nesterov-accelerated Adaptive Moment Estimation.

The hyper-parameters of the chosen optimisation mechanism were left as the default ones [2] as some variations in them led to worsening of the results. The learning rate schedule was changed manually at the beginning of each round. The criterion by which decreases in the learning rate were made consisted in observing the loss on the validation set: where this stabilized, it was decreased at the next round. The decreases in the learning rate were dosed so as not to be too abrupt: every two steps, the learning rate decreases by one decade. In the first round, the initial value of the learning rate is 0.001, in the third it is 0.000316, and in the last it is 0.0001, as is shown in the figure below (Fig. 6).

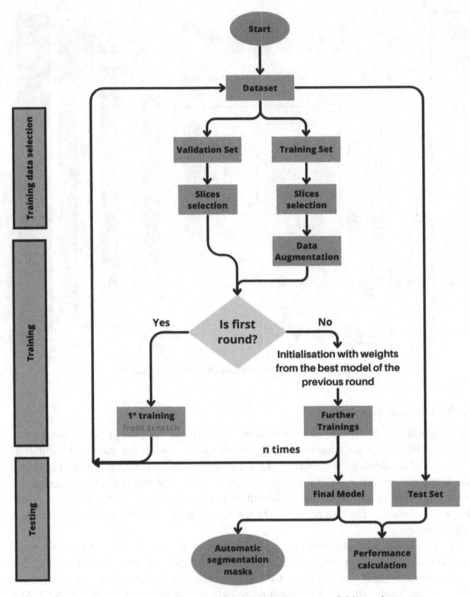

Fig. 5. Summary flow chart of the overall process of this work.

In the following, the trend in loss (Mean Squared Error) during all training rounds (**six training rounds in cascade in the case of the model recounted in this paper**) is shown. The losses calculated on the training set and on the validation set, as the epochs of the various training rounds vary:

Although it is not very noticeable from the graph, the loss values on the validation set are smaller and smaller, confirming that the chosen approach results in a slight improvement of the model as it progresses. Finally, it should be noted that the model

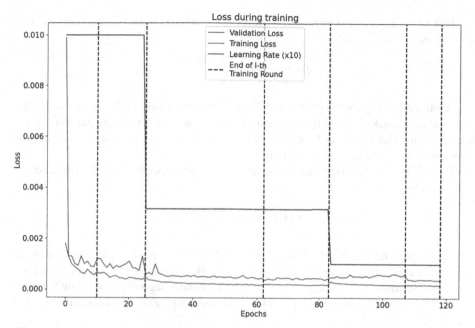

Fig. 6. Loss trend on the training set and validation set during the different training rounds.

that is saved during the training set is the one that has a lower loss on the validation set than the previous models; thus, each successive model is necessarily better than the previous one, without incurring overfitting problems, since the criterion for choosing whether or not to save the model depends on the loss on the validation set, and not on the training set. Therefore, due to the logic of saving, the best model, which was then used to calculate the results of this paper, is the last model of the last round.

3 Results

As can be seen, the training lasted six rounds, each lasting about 10 h, using an A100 GPU, and the inference times of the trained model are about 50 ms per image.

To evaluate the performance of the model, it was decided to adopt two evaluation metrics:

- Dice Similarity Coefficient (DSC)
- Relative Volume Difference (RVD)

These metrics have been implemented for the following class groups:

- Kidney and Masses: Kidney + Tumor + Cyst
- Kidney Mass: Tumor + Cyst
- Tumor: Tumor only

DSC: It is a statistical tool that measures the similarity between two datasets. It is defined as:

$$DSC(X, Y) = \frac{2|X \cap Y|}{|X| + |Y|}$$

In our case X represents the manual 3D mask of the subject in the dataset and Y the automatic 3D mask obtained from the model. The values that this parameter can assume are between 0 and 1. It takes on a value of 1 in the case of a perfect overlap between the two sets and 0 in the case of no overlap.

RVD: This is a statistical tool that measures the difference between two data sets. It is defined as:

$$RVD(X, Y) = \frac{|Y| - |X|}{|X|}$$

where X represents the manual 3D mask of the subject in the dataset and Y the automatic 3D mask automatic mask obtained from the network output. This parameter has only a lower bound equal to -1. A negative value indicates under-segmentation while the more positive the value, the more the algorithm over-segments. Zero represents the optimal situation.

The one giving the best results in terms of loss was chosen as the final model, with which the final performances were calculated. The results shown below were calculated on the entire test and training set (Tables 2 and 3).

Table 2. Metrics that measure the goodness of the segmentation obtained from the model on the whole test set.

Test Set	Mean DSC (%) ± STD	Mean RVD
Kidney + Tumor + Cyst	97.71 ± 1.1	−0.0112 ± 0.017
Tumor + Cyst	81.39 ± 17.4	−0.0215 ± 0.659
Tumor	73.81 ± 24.6	−0.1065 ± 0.491

Table 3. Metrics that measure the goodness of the segmentation obtained from the model on the whole training set.

Training Set	Mean DSC (%) ± STD	Mean RVD
Kidney + Tumor + Cyst	97.80 ± 0.8	0.0040 ± 0.020
Tumor + Cyst	80.98 ± 19.2	−0.0383 ± 0.622
Tumor	80.74 ± 20.4	−0.1503 ± 0.257

Below are some examples of model segmentation on images belonging to the test set (Figs. 7 and 8):

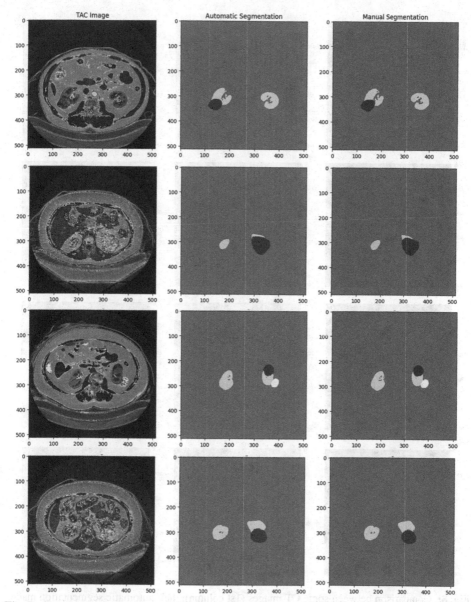

Fig. 7. Comparison between some CT images (1st column), their automatic segmentation made by the model of this work (2nd column) and their manual segmentation (3rd column).

As can be seen, on average, the model fits the actual distribution rather well.

Let's talk about the disadvantages of the model.

As might be expected, the results on the Kidney + Tumor + Cyst complex are better than the others, since kidney is the most represented class. It should be noted that the standard deviations of the Tumor + Cyst and Tumor are particularly high (~20%); this

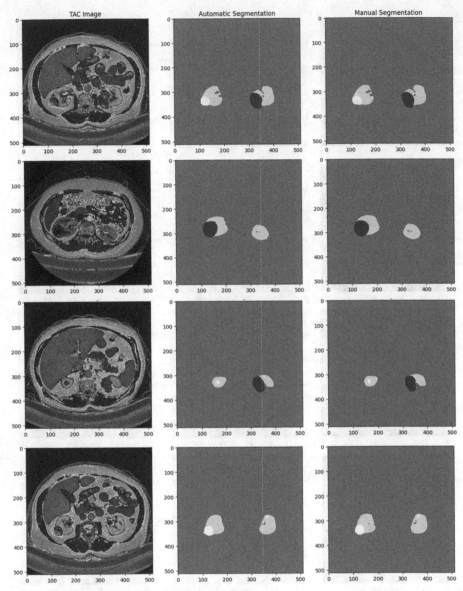

Fig. 8. Comparison between some CT images (1st column), their automatic segmentation made by the model of this work (2nd column) and their manual segmentation (3rd column).

happens because sometimes the model predicts segmentation very accurately, others it does so less accurately. An example that shows this very clearly is the following situation:

In these cases, the model disregards the presence of the tumor. This causes a decrease of the DSC metric, which explains the great variability of the latter in some cases (Fig. 9).

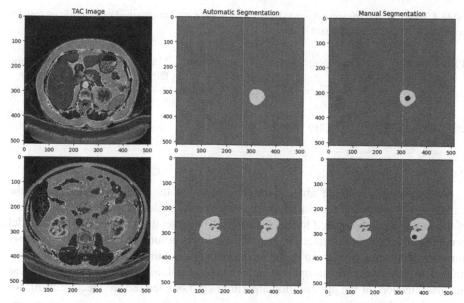

Fig. 9. Drawbacks. Example of how the model sometimes (rarely) loses sight of the tumor.

Another noteworthy aspect concerns the RVD values, which are more variable in the masses case; this means that the model sometimes tends to over-segment and sub-segment approximately symmetrically. In the following is shown a case of over-segment (Fig. 10):

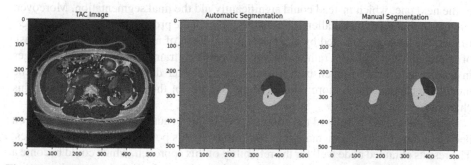

Fig. 10. Drawbacks. Example of how the model sometimes over-segments pathological masses, in this case the tumor.

3.1 Performance on Official Test Set

The final proof that measures the model's actual ability to generalise its task to real data is to run it on an official test set provided by the challenge organisers. The results obtained by the author of this paper, taken from the Leaderboard, are shown in the table below (Table 4).

Table 4. Metrics that measure the goodness of the segmentation obtained on the official test set from MICCAI 2023 Leaderboard.

Place	Team	Average Rank	Dice	Surface Dice	Tumor Dice	Kidney + Masses Dice	Masses Dice	Kidney + Masses SD	Masses SD	Tumor SD
20	Antonio Vispi	20.0	0.719	0.572	0.570	0.941	0.645	0.869	0.453	0.393

As can be seen from the table, all the choices and all the research carried out led the trained model to reach 20th position in the overall ranking, denoting a fair degree of skill of the model when compared to real situations. The generalisation ability of the model is not excellent, albeit discrete. The hypothesised reason for this behaviour is probably attributable to some overfitting of the model with respect to the data with which it was trained, together with the sub-optimal combination of training hyperparameters (such as the loss function, or the optimisation algorithm etc.) or even the architecture itself, for this model and these data.

4 Discussion and Conclusion

One of the major conceptual problems with the proposed model is that it only makes inferences about one image at a time, whereas there are correlations between one image and the next. This limits the skill of the model as it has no idea what is in the previous slice or the next one, which instead could significantly aid the final segmentation. Moreover, another limitation of this challenge lies in the nature of the processed images: CT images. In fact, the latter are affected by a considerable amount of noise, which makes the task of automatic segmentation rather arduous. Probably to train a model with MR images instead, would lead to much better results, as the better definition would enable the models to capture much more intimately the features of the classes of interest in this challenge.

This short paper discussed the possibility of using an automatic approach, based on deep learning, for the segmentation of the kidney and its possible pathological masses. The results obtained bode well for the reliability of this approach, which could become a significant aid to medical personnel, with the aim of increasingly improving the quality and efficiency of diagnosis.

References

1. https://github.com/qubvel/segmentation_models.git
2. https://www.tensorflow.org/api_docs/python/tf/keras/optimizers/experimental/Nadam
3. Zhang, P., Yang, L., Li, D.: EfficientNet-B4-Ranger: a novel method for greenhouse cucumber disease recognition under natural complex environment. Comput. Electron. Agric. (2020)

Attention U-Net for Kidney and Masses

Duho Lee[✉] and Heeyeon Choi

Department of Applied Data Science, Sungkyunkwan University, Seoul, Republic of Korea
dtmanias@gmail.com

Abstract. We proposed new attention method using attention- unet architectures that reflects the 3d axial information to learn the spatial features of 3D images. It showed better performance than baseline unet.

Keywords: Medical Image Segmentation · kits23 dataset

1 Introduction

In the field of 3D medical image segmentation, many methods including U-Net is applied to various applications like cancer diagnosis. Especially, success of image segmentation using 3D CT Image is important to increase the accuracy of cancer diagnosis by assisting the doctors. In many medical image challenges, many attendees used u-net [1] architecture and using nn-unet library. nn-unet is very powerful and showed good results in many applications of 3d Medical Image Segmentation tasks. However, it is hard to modify the architecture and difficult to experiment various settings because it is too "black-box". And attention based architecture showed good performances on many applications like NLP, Computer vision, and also in medical image segmentation. Thus we proposed new attention method using attention- unet architectures [2] that reflects the 3d axial information to learn the spatial features of 3D images. For implementation, we reconstructed the learning architectures based on nn-unet and experimented the proposed attention algorithms. It showed better performance than baseline unet.

2 Methods

2.1 Training and Validation Data

Our submission made use of the official KiTS2023 training set alone.

2.2 Preprocessing

We changed all images into the same resolution because each image has a different resolution. We resample all cases to a common voxel spacing of $1.8410 \times 1.8410 \times 2.3565$, and train the network with a patch size $128 \times 128 \times 128$. To deal with the class imbalance problem, we oversampled the foregroud part which has kidney with probability of 0.5. Also, to improve the diversity of training data and robustness of the trained model, we augmented the training dataset with probability of 0.5, such as Gaussian Noise, GammaTransform, GammaGaussian- blur and so on.

© The Author(s), under exclusive license to Springer Nature Switzerland AG 2024
N. Heller et al. (Eds.): KiTS 2023, LNCS 14540, pp. 139–142, 2024.
https://doi.org/10.1007/978-3-031-54806-2_19

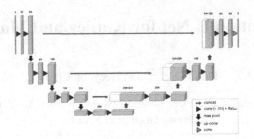

Fig. 1. 3D U-Net architecture and block parameters

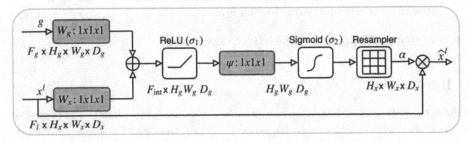

Fig. 2. Attention Gate Block

2.3 Proposed Method

Network Architecture. Our Method is based on 3D U-net [1] and Attention U-net [2], the pipeline consist of two part. First, We use the 3D U-Net as the baseline network architecture, the block information and some parameter information as shown in the Fig. 1. To improve the accuracy of the segmentation result, At first we thought of applying attention to each other in the images. We constructed a model based on the Attention U-net structure as shown in the Fig. 2. Also we assumed that the image should have information corresponding to each axis. We present a new methodology for applying attention based on different axes as shown in the Fig. 3. The axial attention module composed of convolution and batch normalize each axis. in ZX, YZ and ZY axis respectively. and concatenated same demension of input size. These axial attention module blocks implemented in every stages of decoder before applying attention gate block of Attention U-net.

Loss Function and Strategy. We train the model with the combination of dice loss and cross entropy loss. For cross entropy, we multiplied same weights on all classes. The stochastic gradient descent (SGD) algorithm with a momentum of 0.99 was adopted as the optimizer. The learning rate was initialized to 0.01.

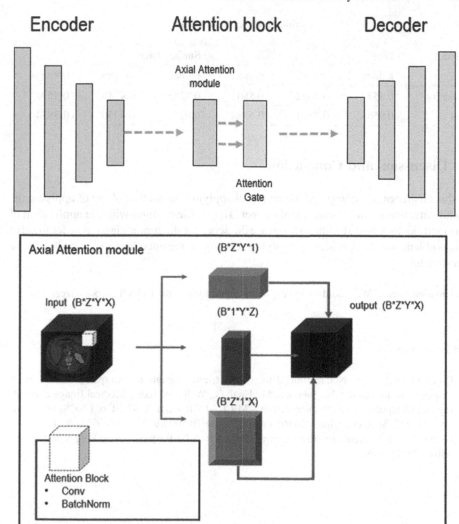

Fig. 3. Pipeline of our proposed method

3   Results

We summarize the quantitative results in Table 1. Dice for kidney, kidney masses, kidney tumor are respectively 0.9565, 0.7074, and 0.6388. Surface Dice for masses, kidney tumor are respectively 0.8240, 0.4350, and 0.3834.

Table 1. Experiment results

Model	Dice			Surface Dice		
	kidney	masses	tumor	kidney	masses	tumor
Baseline	0.9547	0.7092	0.5805	0.8283	0.4544	0.3594
our	0.9565	0.7074	0.6388	0.824	0.4350	0.3834

4 Discussion and Conclusion

In this challenge, we compared the results of applying the methodology of applying the Axial Attention module based on 3D U-net. The problem found with the application of this methodology was still the low surface die score for the tumor class. In order to solve this problem, we are considering applying attention at the encoder level as well as at the decoder level.

Acknowledgment. We would like to express our gratitude to the KiTS2023 organizers.

References

1. Çiçek, Ö., et al.: 3D U-Net: learning dense volumetric segmentation from sparse annotation. In: Ourselin, S., Joskowicz, L., Sabuncu, M., Unal, G., Wells, W. (eds.) Medical Image Computing and Computer-Assisted Intervention – MICCAI 2016. MICCAI 2016. LNCS, vol. 9901, pp. 424–432. Springer, Cham (2016). https://doi.org/10.1007/978-3-319-46723-8_49
2. Oktay, O., et al.: Attention u-net: learning where to look for the pancreas. arXiv preprint arXiv: 1804.03999 (2018)

Advancing Kidney, Kidney Tumor, Cyst Segmentation: A Multi-Planner U-Net Approach for the KiTS23 Challenge

Sumit Pandey[1(✉)], Toshali[3], Mathias Perslev[2], and Erik B. Dam[1]

[1] Department of Computer Science, University of Copenhagen, Copenhagen, Denmark
supa@di.ku.dk
[2] Cerebriu, Copenhagen, Denmark
[3] Copenhagen, Denmark

Abstract. Accurate segmentation of kidney tumors in medical images is crucial for effective treatment planning and patient outcomes prediction. The Kidney and Kidney Tumor Segmentation challenge (KiTS23) serves as a platform for evaluating advanced segmentation methods. In this study, we present our approach utilizing a Multi-Planner U-Net for kidney tumor segmentation. Our method combines the U-Net architecture with multiple image planes to enhance spatial information and improve segmentation accuracy. We employed a 3-fold cross-validation technique on the KiTS23 dataset, evaluating Mean Dice Score, precision, and recall metrics. Results indicate promising performance in segmenting Kidney + Tumor + Cyst and Tumor-only classes, while challenges persist in segmenting Tumor + Cyst cases. Our approach demonstrates potential in kidney tumor segmentation, with room for further refinement to address complex coexisting structures.

Keywords: Multi-Planner U-Net · kidney tumor · segmentation · KiTS23 challenge

1 Introduction

Kidney cancer represents a significant global health burden, affecting a substantial number of individuals and resulting in a considerable number of deaths each year. With kidney tumors being even more prevalent, accurate and efficient radiographic characterization of these tumors is of paramount importance for guiding treatment decisions and predicting patient outcomes. The Kidney and Kidney Tumor Segmentation challenge (KiTS23) serves as a platform for the development and evaluation of state-of-the-art automatic semantic segmentation systems to address this critical need.

According to recent statistics [1], kidney cancer is diagnosed in over 430,000 individuals annually, resulting in approximately 180,000 deaths. Distinguishing between malignant and benign tumors remains a challenge in current radiographic assessments [2]. Moreover, while certain tumors are indolent and exhibit slow growth, the risk of metastatic progression necessitates accurate risk stratification and personalized treatment planning [3]. There is a pressing need for objective and reliable systems capable of

© The Author(s), under exclusive license to Springer Nature Switzerland AG 2024
N. Heller et al. (Eds.): KiTS 2023, LNCS 14540, pp. 143–148, 2024.
https://doi.org/10.1007/978-3-031-54806-2_20

characterizing kidney tumor images to improve risk stratification and predict treatment outcomes.

For nearly five years, the KiTS initiative has cultivated a comprehensive, multi-institutional dataset consisting of hundreds of segmented CT scans featuring kidney tumors, alongside anonymized clinical information for each case [4]. This publicly available dataset has not only served as a benchmark for evaluating 3D semantic segmentation methods [5, 6], but also facilitated translational research in kidney tumor radiomics [7, 8].

KiTS23, the third iteration of the KiTS challenge, builds upon the success of its predecessors by introducing an expanded training set comprising 489 cases and an entirely new and previously unused test set encompassing 110 cases. Furthermore, this year's competition incorporates cases captured during the nephrogenic contrast phase, in addition to the previously utilized late arterial phase. This expansion and diversification of the dataset present a more challenging and clinically relevant context, providing an opportunity to assess the performance of modern approaches.

In this paper, we present our approach utilizing a Multi-Planner U-Net [9] for kidney and kidney tumor segmentation in the KiTS23 challenge. The proposed method leverages the advantages of the U-Net architecture, a popular choice for medical image segmentation tasks. However, we enhance the model by incorporating multiple image planes, capturing complementary spatial information to improve segmentation accuracy. By utilizing both axial and coronal planes, our Multi-Planner U-Net aims to provide a more comprehensive representation of kidney structures and tumors.

2 Methods

The methodology employed in this study can be outlined as follows:

2.1 Training and Validation Dataset

Our submission made use of the official KiTS23 training set alone.

2.2 Pre and Post-Processing

In this step, we took the following steps and during this process, we made sure that the affine of the images and labels remained the same:

1. *Resizing and Normalizing:* To overcome limitations in GPU memory and address inconsistencies in the third dimension of images, the initial step was to resize both the images and labels to dimensions of (256, 256, 256). Furthermore, the pixel values of the images were normalized to fall within a range of 0 to 256.
2. *Label transformation:* For the final data analysis, after completing the training, the data labels were modified as follows: Kidney, Tumor, and Cyst (if available) were transformed into Kidney + Tumor + Cyst, Tumor + Cyst, and Tumor only.

2.3 Proposed Method

MPUnet is based on a modified version of the U-Net architecture, which is a convolutional neural network commonly used for image segmentation. It utilizes a technique called multi-planar training, where the input image is rotated along different axes to generate multiple views of the image as shown in Fig. 1. This allows the model to learn a representation of the 3D image volume and improve generalization.

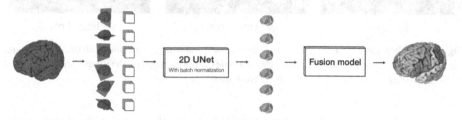

Fig. 1. This figure shown the two-step process of Multi-planner UNet [9]

The system has a fixed model architecture and a fixed set of hyperparameters, eliminating the need for extensive model selection and optimization. It performs well across different segmentation tasks without task-specific modifications, often achieving state-of-the-art performance compared to specialized deep learning methods. The MPUnet framework is open-source and can be used by practitioners without deep learning expertise.

In this study, a robust evaluation of the proposed methodology for kidney and kidney tumor segmentation was conducted using a 3-fold cross-validation technique. The dataset was randomly divided into three subsets: training, testing, and validation datasets. This division was performed three times, ensuring comprehensive coverage of the dataset while minimizing the impact of bias.

During each fold of the cross-validation process, two subsets were utilized for training the model, one for testing the model's performance, and the remaining subset for validation purposes. This approach allowed for an accurate assessment of the methodology's generalization capabilities and ability to handle diverse cases within the dataset.

3 Results and Discussion

In this study (as shown in Fig. 2), we evaluated the performance of a segmentation algorithm on three different classes: Kidney + Tumor + Cyst, Tumor + Cyst, and Tumor only. The evaluation was done using the Mean Dice Score, precision, and recall metrics.

As shown in Table 1, the evaluation of the semantic segmentation model for kidney lesion detection revealed varying levels of performance across different classes. In the training set, the model achieved the highest Mean Dice Score (0.831) in the Kidney + Tumor + Cyst class, indicating accurate segmentation of kidney, tumor, and cyst regions. However, the Mean F1 Score (0.8203) suggested a trade-off between precision

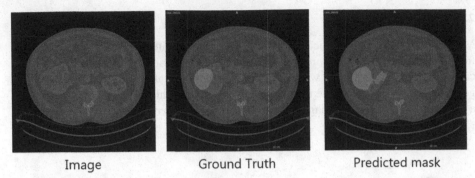

Image   Ground Truth   Predicted mask

Fig. 2. The diagram depicts three components: the original image, a ground truth mask displaying the kidney (in red), tumor (in green), and cyst (in blue) superimposed on the image, and a predicted mask also overlaid on the image. (Color figure online)

and recall. The Tumor + Cyst class exhibited poor performance with a significantly lower Mean Dice Score (0.130) and relatively low Mean Precision (0.564) and Mean F1 Score (0.097), highlighting challenges in accurately segmenting tumors and cysts when present together.

Table 1. Summary of mean dice score, precision, and Recall score for training and testing dataset.

Training set			
Classes	Mean Dice Score	Mean Precision	Mean F1 Score
Kidney + Tumor + Cyst	0.831	0.932	0.8203
Tumor + Cyst	0.130	0.564	0.097
Tumor only	0.26	0.46	0.237
Testing set			
Classes	Mean Dice Score	Mean Precision	Mean Recall
Kidney + Tumor + Cyst	0.832	0.936	0.78
Tumor + Cyst	0.19	0.663	0.14
Tumor only	0.35	0.455	0.36

In the exclusive Tumor class within the training dataset, the model achieved a moderate performance level, as evidenced by a Mean Dice Score of 0.26. This suggests a reasonable ability to detect tumors. A similar pattern was observed in the testing dataset, where the model effectively segmented instances of the Kidney + Tumor + Cyst class, achieving a Mean Dice Score of 0.832, accompanied by strong precision (0.936) and recall (0.78) metrics.

However, when it came to the Tumor + Cyst class, the model's performance was notably lower (Mean Dice Score: 0.19, Mean Precision: 0.663, Mean Recall: 0.14). This indicates challenges in distinguishing between tumors and cysts when they coexist.

Analyzing the challenge results on the Final test set, the model exhibited a Mean Dice Score of 0.816 for the Kidney + Tumor + Cyst class and a Mean Dice Score of 0.364 for masses. By comparing these scores between the training and test sets, it can be deduced that the model didn't suffer from overfitting.

Nonetheless, it's evident that further refinement is required to achieve precise segmentation. While these findings underscore the potential of semantic segmentation models for detecting kidney lesions, the task of accurately segmenting tumors and cysts remains intricate, particularly in cases of coexistence. To enhance segmentation performance, especially for the challenging Tumor + Cyst class—where scores were consistently low across both training and testing sets—additional research and model enhancements are imperative.

4 Conclusion

In this study, we introduced a Multi-Planner U-Net approach for kidney tumor segmentation in the KiTS23 challenge. Leveraging multiple image planes, our method exhibited promising results in segmenting kidney tumors, especially in cases of Kidney + Tumor + Cyst and Tumor-only. However, the segmentation of Tumor + Cyst cases proved challenging, revealing the difficulty of distinguishing between closely associated structures. The cross-validation evaluation highlighted the model's robustness and generalization capabilities. While our approach demonstrates potential, future work should focus on refining segmentation for complex cases, such as Tumor + Cyst, through advanced techniques and additional research. Accurate kidney tumor segmentation holds immense clinical significance, aiding treatment decisions and patient outcomes prediction in kidney cancer management.

References

1. Kidney Cancer Statistics. World Cancer Research Fund. https://www.wcrf.org/dietandcancer/cancer-trends/kidney-cancer-statistics
2. Motzer, R.J., Jonasch, E., Boyle, S., et al.: Kidney cancer, version 4.2017, NCCN clinical practice guidelines in oncology. J. Natl. Compr. Cancer Netw. 15(6), 804–834 (2017). https://doi.org/10.6004/jnccn.2017.0100
3. Patel, H.D., Pierorazio, P.M., Johnson, M.H., et al.: Diagnostic accuracy and risks of biopsy in the diagnosis of a renal mass suspicious for localized renal cell carcinoma: systematic review of the literature. J. Urol. 195(5), 1340–1347 (2016). https://doi.org/10.1016/j.juro.2015.10.180
4. KiTS Challenge: Kidney Tumor Segmentation. https://kits19.grand-challenge.org/
5. Milletari, F., Navab, N., Ahmadi, S.: V-Net: fully convolutional neural networks for volumetric medical image segmentation. In: Proceedings - 2016 4th International Conference on 3D Vision, 3DV 2016, pp. 565–571. IEEE Computer Society (2016)
6. Yu, L., Chen, H., Dou, Q., Qin, J., Heng, P.: 3D deep learning for multi-modal imaging-guided survival time prediction of brain tumor patients. In: Proceedings of the IEEE International Conference on Computer Vision, pp. 1848–1857 (2017)
7. Li, Q., Kim, H., Huang, C., et al.: Generating radiomic features from limited data: a comparative study. Med. Image Anal. 2019(57), 167–178 (2019). https://doi.org/10.1016/j.media.2019.07.008

8. Varghese, B.A., Chen, F., Hwang, D.H., et al.: Decoding tumor phenotypes for renal cell carcinoma using a radiomics approach. AJR Am. J. Roentgenol.Roentgenol. **212**(2), W55–W61 (2019). https://doi.org/10.2214/AJR.18.20497
9. Perslev, M., Dam, E.B., Pai, A., Igel, C.: One network to segment them all: a general, lightweight system for accurate 3D medical image segmentation. In: Shen, D., et al. (eds.) Medical Image Computing and Computer Assisted Intervention – MICCAI 2019. MICCAI 2019. LNCS, vol. 11765, pp. 30–38. Springer, Cham (2019). https://doi.org/10.1007/978-3-030-32245-8_4

3D Segmentation of Kidneys, Kidney Tumors and Cysts on CT Images - KiTS23 Challenge

Marta Kaczmarska[✉] and Karol Majek

Cufix, Grodzisk Mazowiecki, Poland
kaczmarskamarta2@gmail.com

Abstract. Renal tumor, along with renal cyst, is one of the most common kidney diseases. As the kidney tumor incidence is increasing, there is a need for efficient diagnosis and reliable treatment outcomes predictions. Automatic kidney images characterization and differentiating between tumors and cysts could help clinicians with these procedures, providing rapid and repeatable results, free from interobserver variability. The aim of this study is to develop a model for segmentation of kidneys, kidney tumors and cysts on CT scans. For this task we employ a transformer based architecture - Swin UNETR. We conducted a series of experiments to determine which hyperparameters improve the overall model performance measured by Dice score and the model metrics for each class separately. Our best performing model achieves the following Dice scores on test dataset: overall: 51.3, kidney + masses: 78.8, masses: 39.6, tumor: 35.5 and the following Surface Dice scores: overall: 22.6, kidney + masses: 36.7, masses: 16.7, tumor: 14.5. Our model ranked 24th on the leaderboard. The code for our solution is publicly available at https://github.com/deepdrivepl/kits23.

Keywords: Kidney Segmentation · Renal Tumors · Swin UNETR

1 Introduction

Kidney tumor and cyst are two of the most frequent kidney diseases [5]. Once detected, the mass is evaluated to be benign or malignant, which determines the further treatment strategy. The basis for this assessment are Computed Tomography (CT) images [3]. To help with risk assessment and prognosis, a number of nephrometry scores have been developed that use radiomic features [4]. These scores require segmentation of said structures. However, manual annotation is time consuming and expensive and thus, automated methods are preferred. The development of robust kidney, renal tumor and cyst segmentation model would enable fast score evaluation which would improve the efficiency and quality of kidney tumor treatment and surveillance.

© The Author(s), under exclusive license to Springer Nature Switzerland AG 2024
N. Heller et al. (Eds.): KiTS 2023, LNCS 14540, pp. 149–155, 2024.
https://doi.org/10.1007/978-3-031-54806-2_21

2   Methods

2.1   Training and Validation Data

Our submission made use of the official KiTS23 training set alone. We made split on a sorted list of samples by name and used first 80% as training set. The rest 20% of training data were used as a validation set and were not used in training for the final submission.

2.2   Preprocessing

All images' orientation is changed to RAS. The images are resampled to median spacing values of $0.78125 \times 0.78125 \times 3.0$ mm. The HU values are clipped at -200 and 300. To select these values we started with abdomen soft tissue window (W:400 L:50) as a reference and added an offset to account for possible distributional shift (eg. due to tumor subtype). Then, the values are normalized.

For inference the image undergoes foreground cropping after aforementioned preprocessing. The foreground is determined by selecting image values grater than 0.

2.3   Proposed Method

Network Architecture. As a model architecture we chose Swin UNETR [2] implemented in MONAI framework [1]. The model consists of 4 stages with 2 layers on each stage. There are 3, 6, 12 and 24 attention heads at each stage respectively.

Optimization Strategy. We performed a series of experiments and tuned the following hyperparameters:

- patch size: $64 \times 64 \times 64$, $128 \times 128 \times 128$, $256 \times 256 \times 32$, $512 \times 128 \times 32$,
- dimension of network feature size: 12, 24, 36, 48,
- class weights: assigned based on classes' volume percentage or selected by us after exploratory data analysis and first experiments,
- accumulated batch size: 4, 64,
- label smoothing: 0, 0.1,
- number of training epochs: 100, 200, 300, 400,
- learning rates changes according to OneCycle learning rate policy.

Loss Function. For the loss function we used sum of Dice and Cross Entropy loss. We addressed class inbalance while performing patches extractions - drawn patch's center is a specific class based on the chosen classes' ratios. Thus, we did not incorporate class weighting in loss function calculation.

Validation Strategy. We evaluated models performance by calculating overall Dice score and Dice scores for each class separately. We selected only one model with the highest Dice scores to be our final solution. We do not make a models ensemble.

Table 1 shows results on validation set broken down by class for some of our models trained with different hyperparameters listed under *Optimization strategy* subsection. Our best performing model is marked in bold.

Table 1. Models metrics on validation set. Lower indices denote: k - kidney, t - tumor, c - cyst. Default setting: One Cycle Learning Rate and 100 epochs, max LR at 10%, accumulated batch size 4, no class weighting. Abbreviations: f - dimension of network feature size, CW - class weighting, BG - loss including background score calculation, E - epochs, Acc - gradient accumulation after given number of batches.

Architecture	Schedule	Patch size	Dice	$Dice_k$	$Dice_t$	$Dice_c$
SwinUNETR f24	3e-4	64^3	39.3	62.9	13.2	41.8
SwinUNETR f24	3e-4	128^3	42.3	80.3	34.9	11.8
SwinUNETR f24	CW 3e-4	$256^2 \times 32$	41.3	61.8	18.1	43.9
SwinUNETR f24	200E CW 3e-3 BG Acc64	128^3	48.3	73.6	27.5	43.9
SwinUNETR f24	CW 3e-4	$256^2 \times 32$	44.3	66.7	22.2	43.9
SwinUNETR f24	CW 3e-4 BG	$256^2 \times 32$	47.3	72.0	26.0	43.9
SwinUNETR f12	200E CW 3e-3 BG Acc64	128^3	47.4	77.8	28.6	35.7
SwinUNETR f24	200E CW 3e-3 BG Acc64	128^3	46.3	72.6	22.3	43.9
SwinUNETR f36	CW 3e-3 BG Acc64	128^3	46.1	72.1	22.2	43.9
SwinUNETR f48	CW 3e-3 BG Acc64	128^3	43.7	69.8	21.4	39.8
SwinUNETR f12	400E CW 3e-3 BG Acc64	128^3	51.5	76.2	34.3	43.9

3   Results

Our best model takes as an input 3D CT scan divided into smaller patches of size $128 \times 128 \times 128$. During training random patches were extracted from the volume with the following ratios determining the specific class to be the center of the patch: background - 1, kidney - 5, tumor - 10, cyst - 20. During inference sliding window method was used with 0.5 patches overlap and Gaussian-weighted averaging. The model was trained for 400 epochs with RAdam optimizer, weight decay 1×10^{-6} and OneCycle learning rate policy (learning rates: initial 3×10^{-4}, maximum 3×10^{-3} at 20th epoch, final 3×10^{-5}). We performed gradient accumulation every 64 batches. For Cross Entropy loss we used label smoothing 0.1.

The inference of a single 3D image requires 6 GB on a GPU and takes 15 s with sliding window batch size 1 and 0.5 overlap using 16-bit floating-point precision.

Table 2. Challenge specific metrics of our best model on validation and test set.

dataset	Dice ↑				Surface Dice ↑			
	overall	kidney + masses	masses	tumor	overall	kidney + masses	masses	tumor
validation	52.3	85.9	36.6	34.3	31.9	62.6	17.5	15.6
test	**51.3**	**78.8**	**39.6**	**35.5**	**22.6**	**36.7**	**16.7**	**14.5**

Metrics of the best model on validation and test set are shown in the Table 2. These are challenge specific metrics which determine the challenge ranking. We observe a decrease in almost all metrics for the test set. There is a slight increase in Dice for masses and tumor.

An example of our best model predictions is shown in the Fig. 1. It shows potential for improvement, yet the kidney area is correctly determined, thus it can be a promising baseline for further automatic kidney/tumor/cyst segmentation model development.

Figures 2, 3 and 4 show examples of errors the segmentation model makes. Apart from inaccurate objects boundaries, there are difficulties in differentiating between various types of tissues. Small cysts are often omitted. There also occur false positive masses located not in the vicinity of the kidneys, however depending on the target image viewer they may be easily removed during segmentation analysis by the clinician.

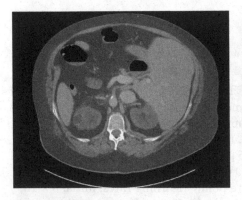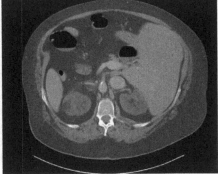

Fig. 1. An example of model prediction shown in axial plane. On the left there are true masks and on the right there are predicted masks overlayed on the image. Colors denote: green - kidney, red - tumor, blue - cyst. It can be seen that cyst is confused with tumor. There is a false positive tumor within the left kidney and small ones above the right kidney. There is also small false positive kidney tissue above the right kidney. The slice comes from image *case_00494*. (Color figure online)

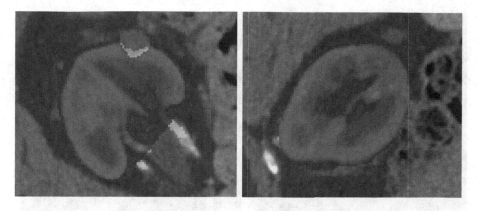

Fig. 2. An example of inaccurate predicted kidney boundaries. The colors denote: green - true positive, red - false negative, yellow - false positive. The false positive region in the upper fragment of the kidney on the left is false negative cyst. The slice comes from image *case_00496*. (Color figure online)

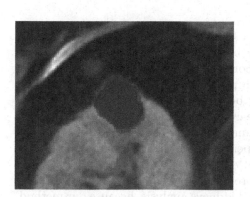

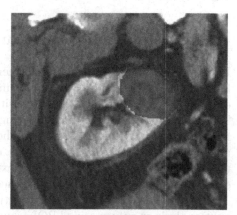

Fig. 3. An example of a cyst mislabeled as a tumor. Green mask is true cyst mask and red mask is predicted tumor mask. The slice comes from image *case_00494*. (Color figure online)

Fig. 4. An example of inaccurate predicted tumor boundaries. The colors denote: green - true positive, red - false negative, yellow - false positive. The slice comes from image *case_00495*. (Color figure online)

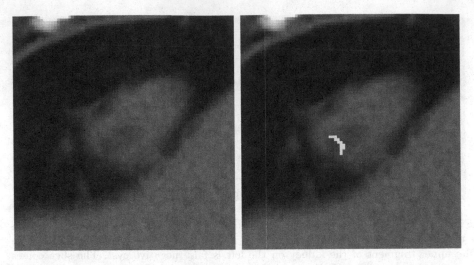

Fig. 5. An example of a doubtful cyst label. The slice comes from image *case_00491*.

4   Discussion and Conclusion

In this study we evaluated transformer based model for the segmentation of kidneys and two of their pathologies: tumors and cysts. The model performs best on kidney, while there is a difficulty with proper tumor and cyst segmentation. There are challenges like differentiating between tumor and cyst and between healthy tissue and masses. Moreover, the cyst is underrepresented in the dataset and occurs only in around half of the images, which contributed to poor cyst tissue recognition by the model. More experiments on the hyperparameter search space for the model should be performed.

It would be valuable to consider postprocessing as well, like connected component analysis or only specified ROI around kidneys analysis, because our method suffers from false positive mass detections located further away from the kidneys. They may be especially numerous when the scan has wide range and apart from abdomen region contains for example legs or lungs. Eliminating such false positives in a postprocessing step could greatly improve our performance metrics.

The performance of the model may be impacted by true labels inaccuracies. We found doubtful labels in the dataset, unfortunately due to lack of a radiologist in the team we are not able to determine them valid or not. Figure 5 shows an example of such label.

The future work could include verifying another promising transformer based architecture - XUnet[1]. Another curious approach to the problem would be pretraining the network in a self-supervised manner on a larger, unlabelled dataset.

---

[1] https://github.com/lucidrains/x-unet.

References

1. Cardoso, M.J., et al.: MONAI: an open-source framework for deep learning in healthcare. arXiv preprint arXiv:2211.02701 (2022)
2. Hatamizadeh, A., Nath, V., Tang, Y., Yang, D., Roth, H., Xu, D.: Swin unetr: swin transformers for semantic segmentation of brain tumors in MRI images (2022)
3. Heller, N., et al.: The state of the art in kidney and kidney tumor segmentation in contrast-enhanced CT imaging: results of the kits19 challenge. Med. Image Anal. **67**, 101821 (2021)
4. Heller, N., et al.: The KITS21 challenge: automatic segmentation of kidneys, renal tumors, and renal cysts in corticomedullary-phase CT (2023)
5. Islam, M., Hasan, M., Hossain, M., Alam, M.G.R., Uddin, M.Z., Soylu, A.: Vision transformer and explainable transfer learning models for auto detection of kidney cyst, stone and tumor from CT-radiography. Sci. Rep. **12**, 11440 (2022). https://doi.org/10.1038/s41598-022-15634-4

Kidney and Kidney Tumor Segmentation via Transfer Learning

Nozadze Giorgi(✉)

Ludwig Maximilian University of Munich, Munich, Germany
G.Nozadze@campus.lmu.de

Abstract. Recently Segment anything model (SAM) has shown great promise for natural image segmentation. This model was trained on by far the largest segmentation dataset, consisting of over 11 million diverse images and 1 billion corresponding masks. The dataset's impressive size and high quality, combined with the powerful Transformer-based architecture, enabled the model to grasp a general understanding of objects and achieve exceptional zero-shot performances, sometimes even outperforming fully supervised models.

However, despite the significant advancements within the zero-shot framework, there are challenges when applying it to more specialized domains like medical and satellite imaging. Due to the scarcity of images from those domains in the training corpus, the model is not as accurate as it could be. Additionally in the fields where the segmentation of only certain, critical areas is desired using the SAM model can be overwhelming.

In this paper, We aim to make use of different Transfer Learning techniques, such as Feature Extraction and Fine-tuning, and investigate different slight adaptations of the architecture to improve the performance of the SAM model and achieve high performance on a given medical image segmentation task.

Keywords: Segment Anything[2] · Semantic Segmentation · Transfer Learning

1 Introduction

Kidney cancer is a prevalent disease affecting a substantial number of individuals worldwide, with more than 430,000 new diagnoses and approximately 180,000 deaths reported annually. Detecting and accurately characterizing kidney tumors presents a significant challenge in clinical practice, as radiographically distinguishing between malignant and benign tumors remains a complex task. For this reason, active surveillance of small renal masses using modern computer vision techniques is becoming increasingly popular by proving its effectiveness.

This paper aims to address the automated semantic segmentation method on the dataset provided by the KiTS23 challenge [1] and existing research to contribute to the improvement of tumor segmentation tasks by utilizing the latest advancements in the field of computer vision.

© The Author(s), under exclusive license to Springer Nature Switzerland AG 2024
N. Heller et al. (Eds.): KiTS 2023, LNCS 14540, pp. 156–162, 2024.
https://doi.org/10.1007/978-3-031-54806-2_22

2 Method

For our project, we have decided to work with the smallest pre-trained model from the Segment Anything Series, specifically the "vit-b" model. The model checkpoint is publicly available in Facebook research's official GitHub repository, allowing us to access its pre-trained weights.

2.1 Model Architecture

The model has the following network architecture:[F1] (Fig. 1).

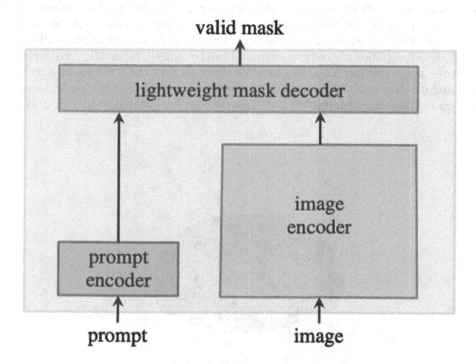

Fig. 1. Model Architecture [2]

In the first stage, it uses an MAE pre-trained Vision Transformer to produce high-quality image embeddings. In the second stage, The embeddings together with the output of the prompt encoder, which encodes the positional information of the desired area on the image are fed to the lightweight mask decoder, which is a modified Transformer decoder block followed by a prediction head, generates the mask prediction.

The SAM model gives an option to incorporate different types of prompts to provide positional and contextual information. It can be prompted using sparse prompts in the form of point coordinates (x, y) or bounding box coordinates (x1, x2, y1, y2) as well as dense mask prompts. In our study, we opted for bounding box prompts because of their flexibility and the clarity of emphasis on areas of interest compared to the potentially ambiguous nature of point prompts.

In a nutshell, given a 3-channel RGB image of shape $3 \times 1024 \times 1024$, the MAE Vision Transformer generates image embeddings of size $256 \times 64 \times 64$. Subsequently, the importance of the regions in the image (and in the feature space) is emphasized and injected using prompt encoder-generated sparse prompt embeddings of size 2×256. These two combined yield a low-resolution prediction mask of size 256×256, which in the final step is post-processed to match the initial input image. After reviewing the zero-shot performance illustrated in Figure **F2**, which was generated using the "Automatic Mask Generation" framework described in the official segment anything paper [2], we believe, that the embeddings generated by the image encoder are of high quality. As a result, we have made the decision to freeze its parameters. This strategic move will prove advantageous as it significantly reduces computational overhead, especially considering that the image encoder alone accounts for about 85 % of the total parameters in the architecture, enabling us to allocate our resources and time efficiently, focusing on re-training or fine-tuning the mask decoder for optimal results (Fig. 2).

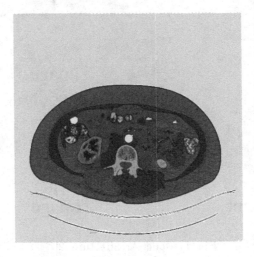

Fig. 2. Zero-shot performance of SAM vit-b

2.2   Training and Validation Data

Our submission made use of the official KiTS23 training set alone. We will use 90% of the available data for training purposes and the remaining 10% will be reserved for validation purposes, allowing us to assess the effectiveness and generalization capabilities of the resulting trained models across different experiments.

2.3   Understanding and Cleaning Data

Before beginning to process the data, as a first step, we analyzed the dataset and found 3 outliers in terms of image resolution (case_00160; case_00419; case_00425) that did not correspond to the 512 × 512 image size. To keep the consistency in the dataset, we reshaped them using bilinear interpolation.

Moreover, we visualized the average pixel distributions of each case and investigated the positive and negative sample ratios across different cases. By conducting this analysis, we gained valuable insights into the class imbalances present in the dataset. These findings played a crucial role in determining the appropriate resampling strategies and selecting suitable loss functions to tackle the class imbalance problem.

2.4   Preprocessing

Our approach consists of two main stages. In the first stage, we generate image embeddings using the image encoder described above. To ensure compatibility with the Vision Transformer, we perform several preprocessing steps on the images: Firstly, we convert the images into RGB format, after that we change the resolution to 1024 × 1024 pixels and finally, we normalize the pixel values of the images and feed them to the image encoder network. The resulting embeddings are then saved on the disk and we move on to the second stage and start training the mask decoder.

2.5   Training

Resampling. In our training process, we handle the batch size as a HyperParameter and are experimenting with different values, whereas the values of 64 or 128 slices have shown very promising results. Every slice within the batch is treated as an independent 2D image. This assumption, although restrictive, is imperative due to the inherent limitation of the SAM model, which can only handle 2D images as input. Additionally, efforts to explore alternative approaches for integrating sagittal or coronal views during training were unsuccessful. Hereby, we will sample batches by shuffling all cases, nevertheless maintaining the positive-negative sample ratio in the batch according to our analysis mentioned in the previous subsection. Moreover, we incorporate several data augmentation techniques to improve the generalization performance of the model. These techniques include rotation, scaling, and contrast adjustment.

Optimizer and Loss Function. We train the mask decoder using the AdamW optimizer with the beta coefficients set to [0.9, 0.999], a weight decay of 0.1, and a dynamic learning rate. Initially, we set a high learning rate of 5e-5 for a specific number of warm-up steps. After the warm-up period, the learning rate is decayed over time using cosine annealing to facilitate effective learning and convergence of the mask decoder. Regarding the Loss Function, we explored various loss functions to address the class imbalance issue. However, as our primary criterion, we selected the weighted sum of Dice and categorical Cross-Entropy loss with class weights, which were set according to our analysis of average pixel distributions mentioned in the subsection Understanding and Cleaning Data.

Models. In total 3 different mask decoders were trained.

1. **ROI Decoder** was trained using a fixed, large box prompt[F3a] to identify regions of interest (Kidneys) in the given 2D slice. The binary masks produced by this model aid in the creation of prompts for the second phase of training, geared towards identifying cysts and tumors. To ensure comprehensive coverage of the complete kidney area, the bounding boxes drawn around the binary mask instances are intentionally expanded by incorporating a slight random factor.[F3b]
2. **Tumor/Cyst Decoders** were trained separately due to SAM's difficulty in identifying multiple objects within a single prompt framework. This issue will be further discussed, along with other relevant challenges and limitations, within the upcoming "Discussion and Conclusion" section (Fig. 3).

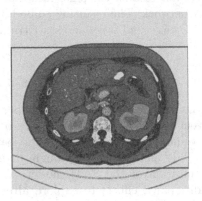
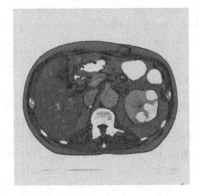

(a) Prompt used in training of ROI Decoder

(b) Prompt used in training of Tumor/Cyst Decoders

Fig. 3. Overview of Prompts in different decoders.

Resources. The training process of a single mask-decoder was carried out on a single GPU with 40 GB of GPU memory, spanning a total duration of two days (Fig. 4 and Table 1).

3   Results

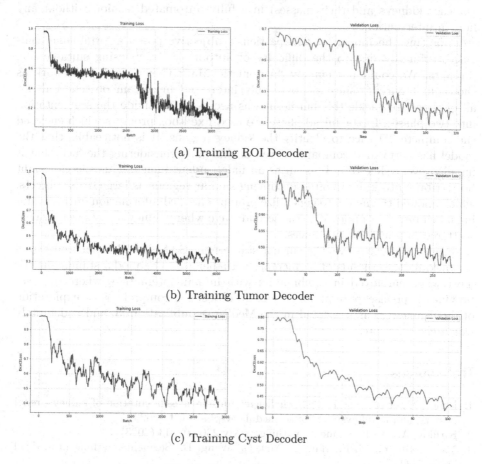

(a) Training ROI Decoder

(b) Training Tumor Decoder

(c) Training Cyst Decoder

Fig. 4. Training Overview

Table 1. Results on the official test set

Dice	Surface Dice	Tumor Dice	Kidney+Masses Dice	Masses Dice
0.432	0.239	0.222	0.807	0.268

For a detailed overview of the training results, including the model checkpoint and an example notebook, please refer to the "TransferSAM" repository [4].

4   Discussion and Conclusion

Back in April, when Meta AI introduced the SAM model, discussions began to circulate about its exceptional performance in segmenting natural images. This paper and the method it proposes serve as an experiment to test the model's effectiveness in a scenario where the goal is to solely identify certain objects (in our case kidneys and their masses) in a fully automated fashion without any human intervention.

Adapting the model for this particular objective posed a formidable challenge, primarily due to the difficulty of automatically choosing suitable box prompts. We could not employ "Automatic Mask Generation" framework as shown in Figure[F2], since we were only interested in certain objects and not all of them. To tackle this hurdle, it was necessary to divide the segmentation into two phases. In the initial phase, we used fixed-box prompts which enclosed the complete 2D slice to identify the kidney regions. It is noticeable, that the model has performed comparably well in this part, considering the fact, that it handled every slice of a 3D scan as an independent image. This success might be attributed to kidneys often occupying similar regions as large dense objects, which allowed the model to generalize prompt encoded information during training much better as opposed to the second stage where kidney masses were widely scattered into multiple instances.

From my perspective, the current state of the SAM model is not well-suited for tackling these kinds of segmentation challenges. The model stands out the most when employed in combination with human interaction, where the user provides a precise prompt for a specific object. A comprehensive exploration of this framework is presented in the MedSAM publication [3] and is definitely worth checking out.

References

1. Heller, N., et al.: The KITS21 challenge: automatic segmentation of kidneys, renal tumors, and renal cysts in corticomedullary-phase CT (2023)
2. Kirillov, A., et al.: Segment anything. arXiv:2304.02643 (2023)
3. Ma, J., He, Y., Li, F., Han, L., You, C., Wang, B.: Segment anything in medical images (2023)
4. Nozadze, G.: Transfersam (2023). https://github.com/Noza23/TransferSAM.git

Author Index

© The Editor(s) (if applicable) and The Author(s), under exclusive license
to Springer Nature Switzerland AG 2024
N. Heller et al. (Eds.): KiTS 2023, LNCS 14540, pp. 163–164, 2024.
https://doi.org/10.1007/978-3-031-54806-2

Printed in the United States
by Baker & Taylor Publisher Services

Printed in the United States
by Baker & Taylor Publisher Services